Lecture Notes in Medical Informatics

Edited by P. L. Reichertz and D. A. B. Lindberg

34

R. Janßen G. Opelz (Eds.)

Acquisition, Analysis and Use of Clinical Transplant Data

Proceedings

Springer-Verlag Berlin Heidelberg GmbH

Editors

Dr. Rainer Janßen
IBM Deutschland GmbH, Wiss. Zentrum Heidelberg
Tiergartenstraße 15, 6900 Heidelberg, FRG

Professor Dr. Gerhard Opelz
Institut für Immunologie der Universität Heidelberg
Im Neuenheimer Feld 305, 6900 Heidelberg, FRG

ISBN 978-3-540-18511-6   ISBN 978-3-642-51003-8 (eBook)
DOI 10.1007/978-3-642-51003-8

© Springer-Verlag Berlin Heidelberg 1987
Originally published by Springer-Verlag Berlin Heidelberg New York in 1987

Contents

List of Contributors

P.G. Beatty

P. Conort

E. Donnall Thomas

R. Engelbrecht

H.P. Friedman

J.A. Hansen

R. Janßen

E. Keppel

H. Lange

C. Lohrengel

P.J. Martin

M.R. Mickey

H.-G. Müller

T. Müller

G. Opelz

R. Reuter

P. Roebruck

N.H. Selwood

J. Sequeira

P.I. Terasaki

K. Ullmann

The international symposium on Acquisition, Analysis and Use of Clinical Transplant Data, held in Heidelberg, October 29-31, 1986, brought together physicians, statisticians and computer scientists from 12 countries in Europe and North America. The intention was to provide an interdisciplinary forum for discussion of current issues and trends in clinical transplant research.

Transplantation has become the accepted method of choice for the treatment of human organ failure, especially kidneys. With ever increasing knowledge of the body's immune system and better immunosuppressive drugs, hearts, livers and bone-marrow are transplanted at rapidly growing rates.

The genetic system steering the reaction of the human body against foreign tissue is extremely complex. Therefore, international collaboration is essential in acquiring and distributing donor organs to suitable recipients as well as in research aimed at a better understanding of pre- and post-transplant treatment, patient selection, and so forth. The need for collaboration led some 20 years ago to the foundation of institutions like EUROTRANSPLANT, the European Dialysis and Transplant Association (EDTA), and similar organisations in other parts of the world.

In 1981, an international study was initiated by G. Opelz, Department of Transplantation Immunology at the University of Heidelberg, to centrally collect historical data from patients who received a kidney transplant. In the frame of this Collaborative Transplant Study (CTS), over 250 transplant centers in 34 countries submit regularly a set of predefined patient data. More than 40,000 kidney transplants have been recorded thus far. This constitutes the most extensive data collection on transplant case histories in the world.

The international success of the study led to a rapid growth of the amount of data to be processed. The large data collection in Heidelberg, along with growing communication problems of the CTS participants, motivated the University of Heidelberg and the Science Center of IBM Germany at Heidelberg to embark upon a joint research project, called TRAINS (Transplant Information System), to extend the scope of CTS toward new directions. Based on the experience of CTS, the objectives of TRAINS were to explore the potential of modern computer and communication technology in order to improve the cooperation in transplantation immunology between geographically distant transplant centers. Specific problems were :

- The requirements which a central database management system, dedicated to the specific needs of transplant immunology, has to satisfy in order to properly collect, manage, and evaluate the CTS date in a central location.

- The development of tools and procedures for fast and reliable information exchange between the central system and the remote clinics.

- The provision of tools helping researchers to gain useful knowledge from the CTS data.

- To provide an answer to the question 'If a kidney with specific characteristics becomes available for transplantation, where is the optimal candidate on a list of potential recipients ?'

The idea and the concept for the symposium on Acquisition, Analysis, and Use of Clinical Transplant Data originated in this joint research project. It was also our contribution to the list of events celebrating the sixth-centenary of the Heidelberg University. The symposium was structured in three sessions. The first session was dedicated to the design of clinical transplant trials and the analysis of data. In a second session, questions concerning the design and implementation of medical information systems for transplant research were discussed. The last session covered the assignment problem (donor/recipient matching) and related questions.

The symposium was organized by the Scientific Programs Department of IBM Germany and sponsored by IBM Germany. The editors would like to express their gratitude and appreciation to the sponsors, to all lecturers, and to the many contributors within and outside IBM who gave advice and assistance in preparing, organizing, and running this symposium.

Heidelberg, May 1987.

G. Opelz   R. Janßen

DESIGN OF CLINICAL TRANSPLANT TRIALS

P. Roebruck
Institut für Medizinische Dokumentation,
Statistik und Datenverarbeitung
der Universität Heidelberg
Heidelberg, FRG

1. Introduction

Writing or speaking about the methodology of clinical trials is quite a simple business if one is confined to some general aspects. Going into the details of special fields of applications on the whole is troublesome to both the writer and the reader. A very good way to write a paper on clinical transplant trials would be to report an example of a concrete trial which has been designed, conducted, and analysed with the author as the cooperating biometrician. Nevertheless, due to the lack of the author's experience with concrete transplant trials the following representation will settle somewhere between methods and applications, and emphasis will be a little bit more on the methods. Starting with aspects of clinical studies it passes to the question, what type of problems - regarding transplantation - could be attacked experimentally, what means in a trial instead of an observational study. Next, some essential features of clinical trials will be discussed in general and with respect to the experimental-type problems separated before. A short section is dedicated to the ethical problem oftenly associated with clinical experiments. The paper will be finished by a very short remark on observational studies, which clearly may not be excluded from clinical research.

2. Clinical studies

We will begin with the attempt of giving a rough characterization
of clinical studies by their main criteria. These are the questions to
be answered and the methods for doing so.

Some typical problems that may be solved by means of clinical
studies are
- prove of a beneficial influence on the course of a disease by some
  therapy
- comparison of several treatments with respect to their
  effectiveness
- investigation of dose-response relations
- comparison of (sub-)groups of patients with respect to the effectiveness of one treatment
- evaluation of the effect of combining therapies
- prove of the constance of treatment effects in recurring
  applications
- prediction of the course of a disease (under distinct therapies)

Roughly speaking, a clinical study aims at the comparison of
several groups either with respect to a great number of observations
achieved from each person (which could be called an explorative
approach) or with a small number of well defined criteria in a
confirmative approach. Most clinical investigations combine both
elements and in particular it is the latter, which claims a careful
design of the study.

The conclusion, that the criterion is affected by the factor under
consideration (such as "group" or "treatment" or "recurrence") must
rest on an assessment of significance and a judgement of causation.
SCHWARTZ et al. write in their very instructive book (which will be
cited here repeatedly) in the context of evaluating the differences
obtained by two treatments:

"The assessment of significance permits us to say, whether the
observed difference may be merely the result of sampling variability
or is evidence of a real difference. It will be based on a statistical
test.
If the effect is significant, a judgement of causation allows us

to attribute it to the difference between the treatments. This ist only possible, if the groups are strictly comparable in all respects apart from the treatments given."

They add with respect to the latter point: "Providing two such comparable groups is another standard statistical problem the correct solution of which is obtained by randomization". This statement will engage our attention later on.

Thus having turned to methodology, we first will consider some characteristics of clinical studies from that point of view.

2.1 Experimental and nonexperimental studies

In general, one has to differentiate between experimental clinical studies, so called clinical trials, which are the subject of this paper and nonexperimental ones, the observational studies. In a clinical trial some factor of possible influence - like treatment - is allocated to patients by the clinician according to a rule which is established in advance. This rule does not rest on the patients needs alone but also on the needs of the study which in turn is designed for the benefit of future patients. In an observational study however, each patient is treated in an individual way which is thought to be best for him by the physician. In both situations, lots of items are recorded from each patient which from a methodical point of view can be classified as influence factors or target variables or possibly both when considered as a link in a chain of casualties. Thus even in a clinical trial one is concerned with factors which, in part, can not be controlled in an experimental way. The influence of these factors may be known, in this case they are used for stratification in advance, especially to balance sample sizes with respect to them, or unknown, which gives rise for an explorative analysis in addition to the confirmative analysis of the results of the trial. Furthermore, confirmative analysis is sometimes restricted to very few target variables for statistical reasons which appear samewhat technical and therefore are not discussed here.

2.2 Clinical studies and kidney transplantation

To be concrete, let us have a look at some exemplary problems in the framework of kidney transplantation, which surely raise with respect to other allografts in a similar manner. It is suggestive to subdivide them roughly by means of time relative to the operation and whether they primarily relate to the donor and the kidney or to the recipient. Surely, this scheme covers not all variables of possible influence and not all outcomes, which may be of interest. I think however, it is sufficient for a discussion of methodical aspects.

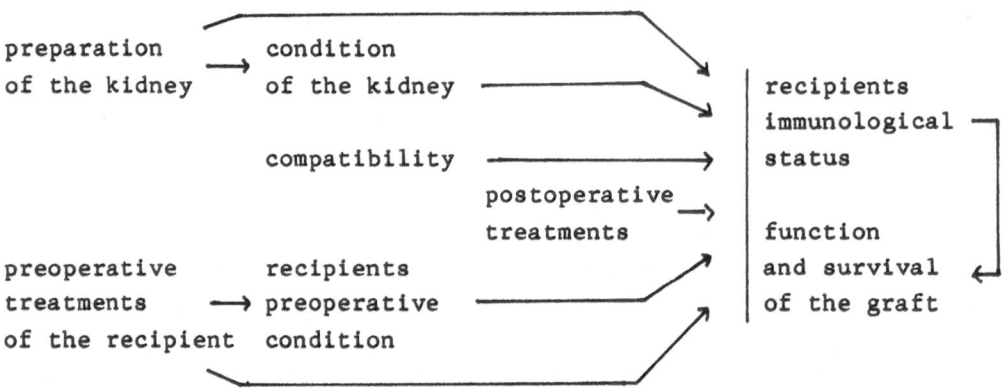

All questions related to the surgery itself are omitted here for simplicity and because it seems to be a routine which is very well done in general.

A few examples of variables which usualy are considered with respect to specified categories are:

preoperative treatment:
dialysis duration
transfusion
prevention of sensitization
plasma exchange

recipient's preoperative
condition:
number of previous transplantations
previous pregnancies
immunological status

preperation of the kidney:
transport
ischemia times
conservation

condition of the kidney: parameters from biopsy and inspection
compatibility: degree of HLA-compatibility
                             blood groups
                             donnor-recipient crossmatching
postoperative treatment: immunosuppression

One factor that clearly must be considered in addition is the center effect (Cicciarelli et al. 1985).

Having in mind the design of clinical trials, we see different types of factors:

- Variables which - despite of ethical arguments perhaps - can actually be controlled in an experiment (like transfusions, prevention of sensitization, immunosuppression).
- Variables which can not be controlled but which should be used for definition of the study population or for stratification either in subset analysis or in prediction (e.g.: dialysis duration, immunological status).
- Variables with a theoretical possibility of being controlled as compatibility, preparation and condition of the kidney. Practically, even with a tremendous effort in conducting the trial, only very small simple sizes could be achieved. In most cases these variables have to be handled as non controllable for the purpose of a trial.

In the following we deal with variables of the first type as subjects of clinical trials. As mentioned above, the other ones or some of them have to be used for stratification in randomization and in the analysis of the treatment results as well as for an explorative analysis of their own effects. Here, we will not stress this last point furthermore.

3. Clinical trials

The terminology for the discrimination of clinical studies is not definit in all aspects. Here we use the following hierarchical classification (Mau et al. 1985).

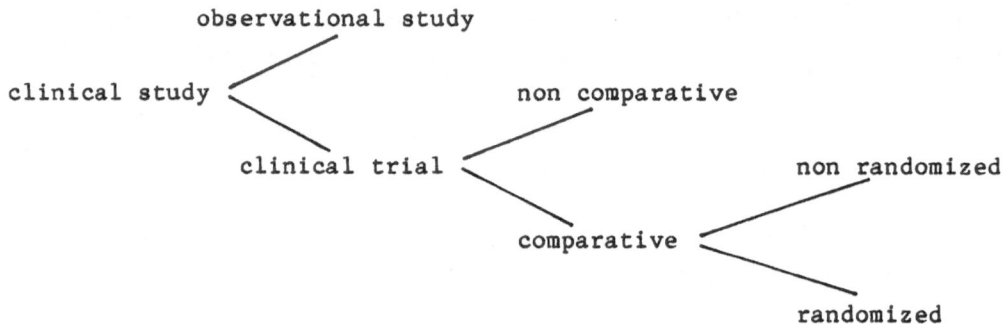

In general, speaking about controlled clinical trials, one has in mind the lowest branch of this tree, which characterizes an ideal type of clinical studies.

3.1 Controlled clinical trials

Essential elements of a controlled clinical trial are

- well defined subjects for the trial
- well defined treatments
- well defined assessment criteria
- randomization
- blindness
- methods for comparison
- predetermined number of subjects

According to Schwartz/Lellouch (1967) we differentiate an explanatory and a pragmatic approach of therapy studies, but we concede, that the approach of many studies lies between these extremes. Loosely speaking, the object of the pragmatic approach is the choice of the best therapy (which may consist in a scheme of several treatments), whereas the explanatory approach aims at the effect of a special part of the therapy or at the prove of a particular mechanism. The choice of some of the elements of a trial depends on the type of the approach. The following comments at the topics listed above follow Jesdinsky (1986) and Schwartz et al. (1980).

Subjects for a trial: The choice of patients for a trial with
a pragmatic approach is related to the question to what patients are
the results of the trial destinated to be applied? With an explanatory
approach the type of patient may be specially chosen to be that which
will best display a well defined biological phenomenon. Even with a
pragmatic approach there remain two problems of extrapolation: from
the patients of the centers conducting the trial to that of other
centers and from the present patients to those in the future.

Treatments: The definitions of the treatments to be compared has
to comprise the treatments themselves, the way they should be
administered and the admissibility of concomitant treatments. From the
explanatory point of view, these definitions should be as narrow as
possible, while a pragmatic approach allows or even demands greater
variability in every respect.

Assessment criteria: "With an explanatory approach the hypothesis
to be tested will concern a simple pathological state. No doubt we can
take adventage of the trial to study other criteria which may suggest
us new hypotheses for investigation but these will be secondary. In
general, the choice of a small number of criteria is the mark of a
clearly formulated hypothesis. ... With a pragmatic approach the
situation is quite different. Here we must take account of all the
practically important criteria, and there will be well many of these.
However, at the analysis stage they cannot be considered singly, for
only one decision can be taken and this must rest upon an overall
balance of advantages and disadvantages." *)

Randomization: "Allocating the patients to treatments by
randomization produces (almost surely) groups of patients which are as
alike as possible with respect to all their characteristics, both
known and unknown." *) Furthermore randomization ensures that a
statistical test for the assessment criterion is meaningful and its
result can be interpreted. Without randomization this only would be
possible by the assumption that the patients in each group of the
trial constitute a random sample from a population of interest. This
is a quite obscure assumption in most clinical situations. Ethical
problems with randomization will be picked up later on.

*) Quotations from Schwartz et al. (1980)

Blindness: With an explanatory approach it is desirable that patients in a trial do not know, what treatment is administered to them. From a pragmatic point of view, blindness of the patients seems to be less important. Furthermore, there are lots of situations where blindness of the patients impossibly can be realised. The doctor's blindness can be useful - although not justifiable sometimes with both approaches - to voide that suggestive effects are different with the different treatments or that the assessment criteria are rated in a manner depending someway on the treatments. A possibility to evade the last pitfall is to get ratings of the nonobjective criteria by a third physician.

Methods for comparison: The usual method to compare treatments with respect to a criterion is a statistical test to decide whether or not the observed differences are due to chance. At least two kinds of error may occur: treatment differences are due to chance but they are judged to be not (with probability α) or vice versa (with probability β. Both error probabilities should be small with respect to an explanatory approach. With a pragmatic approach it may be possible sometimes to use a statistical selection procedure instead of a statistical test. Such procedures select the best of the treatments with respect to the assessment criteria. They make sure that the probability of an incorrect selection does not exceed some prescribed β . In selection procedures the "significance level" is meaningless due to the convention that each of two equivalent treatments is "best". (In fact, for two treatments it can be shown that a selection procedure can be regarded as a test with $\alpha = 0.5$, see Schwartz et al. 1980).

Predetermined number of subjects: A crucial question when planning a trial is, how many patients are needed. The answer depends on the statistical type of the assessment criteria, on the minimal order Δ of a change in the criteria which is considered to be relevant, and on the probability β not to detect a change of this order. Furthermore, it depends on the magnitude of α , when we are concerned with statistical tests. All these have to be fixed in advance of the trial in order to estimate the sample size needed. In general, a selection procedure does with much less patients than a test (with the same values for Δ and β). This advantage however may be compensated in the framework of a pragmatic approach due to the greater variability of the study subjects compared to those in trial with an explanatory approach.

3.2 Transplant trials

As we have ascertained in a previous section, for the example of
kidney transplantation only few of the relevant problems in this area
may be candidates for being attacked by clinical trials. We mentioned
preparing the transplantation by transfusions, eventually escorted by
immunosuppression and the immunosuppressive protocol following
surgery. In particular with respect to the latter it is a widespread
use to compare observations from different time periods: Results from
immunosuppressive protocols using cyclosporine are compared with
"historical data" obtained from a conservative protocol administered
at a time when cyclosporine was not yet available. Perhaps one could
argue that parts of the benefits ascribed to cyclosporine in fact are
due to a general clinical progression or other time dependent factors?

Nevertheless, we will tackle the question whether or not clinical
trials may be a possible method in the field of transplantations in a
constructive manner. To do so, we can look at the elements of a
controlled clinical trial with respect to the three concrete
questions which we have just mentioned.

With the choice of the study population, there will be few
problems regarding the pragmatic approach. One could, for instance,
take patients of medium age from one or from several centers, which
are designated for getting a cadaveric graft, to study the best number
and kind of non-doner-specific transfusions. Clearly, subjects must
be selected so that no individual features preclude the treatment.
Being not an expert in immunologics I would like to avoid offering
subpopulations for the study of mechanisms with an explanatory
approach.

The same holds true regarding the definition of special treatments
or complex therapies. What I should mention here instead, is the
possibility of avoiding ethical problems occasionally by the choice of
treatment strategies connected with suitable criteria for comparing
them. Within the context of immunosuppressive treatments following
surgery one could think of a criterion for monotoring the individual
response to the treatment administered first according to a random
allocation. At certain events - equally defined for all patients -
standing for a bad response, the treatment is changed. To assess the
difference between treatments the time from transplantation until the

change of treatments could be used. For those patients for whom the event never occurs and treatment is not changed, it can be defined as the graft survival time.

In general, survival time or graft survival time, as considered here, is quite an objective criterion and demands not necessarily the blindness of the doctor. The decision whether to change the treatment, however, should be made by a blind doctor or, if this is impossible because of other aspects of patient monitoring, by a third physician, who is unable to relate his observations to the treatments. The same may be true with respect to the evaluation of a patient's immunological status after administering therapies (or placebos) to prevent sensitization resulting from transfusions (e.g. Raftery et al. in Opelz (1985)).

Both types of studies probably could be conducted with blind patients. In trials like those comparing different numbers of transfusions e.g., blindness of patients, however, would demand dummy transfusions to equalize the number of actions for all patients. This seems not justifiable for ethical reasons. With a pragmatic approach it is even absurd, whereas with an explanatory approach it would be desirable for the purpose not to confound differences due to the treatments with those due to suggestion, the so-called placebo effect. A third type of blindness is that of the statistician, who is concerned with the analysis of the study results. We should think over this aspect.

The choice of a method of comparison - test or selection procedure - associated with that of the sample size is not as easy as it may seem at a first glance. The idea not to control the probability of a "wrong" decision when treatments are equal or nearly equal with respect to the criterion looses somewhat of its persuasiveness, if e.g. treatments possibly differ with respect to late side effects which are not recorded in the criterion. Thus, in comparing immunosuppressive monitorings following transplantation, one would prefer a test to assess significance, whereas in presurgery treatments the use of selection procedures may be considerable. In any case, the decision must depend on the concrete question to be investigated, which clearly holds true for all the other aspects of a clinical trial, too.

One of the most important aspects of the decision whether to
conduct a clinical trial or not, has been omitted so far, namly that
of ethics.

3.3 Ethical problems

The few examples, which hopefully have been presented with the
necessary caution, show that the controlled clinical trial may be a
tool to answer some questions in the field of organ transplantations.
What remains is the question, whether an allocation of patients to
treatment is justifiable, which does not depend on an individual
indication or the patients individual preference. The so-called
"informed consent" of patients solves this problem only in part,
because it does not relax the doctor's responsibility. The decision to
conduct a trial has to depend on the concrete circumstances.

I will content myself with the quotation of a general remark of
Schwartz et al. (1980) regarding the example of a promising new
treatment: "If a single doctor, confined as he may be that his new
treatment is better than the standard one, cannot produce a strict
justification for his views, and if other doctors simply do not agree
with him, then a trial is justified. The ethical decision is here at
the level of the medical community. If the trial confirms the settled
opinions of those doctors, who are in favour of the new treatment, the
burden of experimentation may seem heavy to them and they may feel
that they have been asked to sacrifice some of their patients to the
demand of scientific rigour. But this is merely to ignore the
possibility that they might have been wrong. There are plenty of
instances in which a new treatment has proved not to be better than an
old one; in not a few, the new treatment has turned out to be worse
than the old. ... Considerations at group level also relate to the
patient. If we agree that the study of a set of individuals ought to
lead to the establishment of therapeutic principles applicable to
whole groups of patients we are bound to adopt the only methodology
capable of attaining this end. ... Doing an inadequate trial amounts
to taking risks with no chance of benefit. It is this last possibility
which is genuinely unethical."

4. Observational studies

With the last two sentences Schwartz et al. refers to the
possibility to content oneself with observational studies if a trial
cannot be conducted adequately. The gain in knowledge, however, will
be smaller with both an explanatory and a pragmatic approach. In
general, many nonexperimental studies are necessary in order to make
plausible a causal associaton between factors and effects, whereas a
small number of trials could be sufficient. Combining the results from
various observational studies we may follow the principles of the
International Agency for the Research on Cancer for the evaluation of
causalities of associations from epidemiological studies:
"There is no identified bias which would explain the association, the
possibility of confounding has been considered and ruled out as
explaining the association, and the association is unlikely to be due
to chance. In general, also a single study may be indicative for a
cause-effect relationship, confidence in inferring a causal
association is increased, when several independent studies are
concordant in showing the association, when the association is strong,
or when there ist a dose-response relationship."

I like to conclude by mentioning that these principles
also hold true in a certain way in the context of clinical trials.

5. References:

BIEFANG,S./KÖPCKE,W./SCHREIBER,M.A.: Manual für die Planung und
    Durchführung von Therapiestudien, Springer, Berlin 1979

CICCIARELLI,J./MICKEY,M.R./TERASAKI,P.I.: Center effect and kidney
    graft survival, in OPELZ (1985) 2803-7

JESDINSKY,H.J. (ed.): Memorandum zur Planung und Durchführung kon-
    trollierter klinischer Therapiestudien, GMDS-Schriftenreihe 1,
    Schattauer, Stuttgart 1978

JESDINSKY,H.J.: Statistik und Therapieforschung, Münchener Medizini-
    sche Wochenschrift 128 (1986) 701-4

MAU,J./NETTER,P./NOWAK,H./VOLLMAR,J.: Biometrische Aspekte der
    Planung und Durchführung nicht-randomisierter vergleichender
    klinischer Prüfungen, Unpublished Mimeograph 1985

OPELZ,G. (ed.): Relevant immunological factors in clinical kidney
    transplantation, Transplant. Proc. 17 (1985), No. 6, 2175-2827

RAFTERY,M.J. et al.: Prevention of sensitization resulting from
    third-party transfusion, in OPELZ (1985) 2499-2500

SCHWARTZ,D./LELLOUCH,J.: Explanatory and pragmatic attitudes in
    clinical trials, J. chron. Des. 20 (1967) 637-48

SCHWARTZ,D./FLAMANT,R./LELLOUCH,J.: Clinical trials, Academic Press,
    London 1980
    (Translation of "L'essai therapeutique chez l'homme" 1970)

STATISTICAL MODELING FOR KIDNEY TRANSPLANT DATA

M. R. MICKEY AND P. I. TERASAKI

UCLA TISSUE TYPING LABORATORY

The ultimate goal of a mathematical/statistical model is to give a complete quantitative expression of the relation between variables describing outcome and antecedent, conditioning variables. One would like to express graft survival, for example, in terms of the multitude of variables that jointly determine how long and how well a given graft will function. The model could be expressed by equations, graphs, tables or combinations of these modes. The ideal is probably unattainable, and considerable simplification is necessary, particularly considering that each transplant/potential transplant is unique.

It is too much to ask that such models be developed solely on theoretical grounds. We could hardly expect, at present, for example, to predict the quantitative effect of cyclosporine solely on the basis of its chemical description. Consequently the models must have an empirical content. The statistical approach is to state the model in terms of both the givens- number of pre-transplant transfusions, extent of histocompatibility mismatching, donor/recipient ages, etc. - and unknown constants. The values of the constants are then estimated so that the fitted values ("predicted values") correspond to the observed data. Statistical considerations are then concerned with questions related to the appropriateness of the model form, whether or not certain variables should be included, how sensitive the model is to the parameters, how well the constants are estimated, etc.

Statistical concepts also enter into the basic formulation of the model. One would not expect to predict the exact number of days a graft would function at some given level. The model, then, is to express the relation between the _probability_ of the outcome and the conditioning variables. Formally, we require a description of P(Y|x) where x means the constellation of values of the corresponding several conditioning variables.

The vague generality becomes more specific if we make the restricting assumption that outcome is to be assessed in terms of graft survival time. We will suppose that the duration of graft function is defined sufficiently well for all practical purposes. The relevant probability is then the survival function, the probability that the graft will function at least t time units. We express this formally as S(t|x), where again x denotes the values for a set of conditioning variables.

Since the variety of models is large, we need a classification. The simplest class is that of the _additive_ models. Most models we will be concerned with here fall into this class. The general idea is that the influences of the various variables enter the model independently of one another and in an additive manner. The simplest case is that in which the factors are additive on a linear scale, as indicated in figure 1. We take outcome as the per cent one year graft survival. The model expresses the outcome as a summation of scores corresponding to the levels of the variables. In the illustration per cent one year survival is indicated as a constant plus a transplant center effect plus a transfusion effect plus a matching effect. If center 2 and center 10 each transplant with 5 transfusions and 1 mismatch, center 2 will have 60+4+10+2=76 per cent one year graft survival and center 10 will have 60-15+10+2=57 per cent survival. The difference, 19, is the difference of the "center effects", so that the difference is the same regardless of the other factors so long as we consider transplants that are the same for the other factors. We could say then that the center effect is independent of the other factors.

This simple additive model is certainly convenient. The problems are: how are the effects for the factors to be determined? how appropriate is the model? We will defer these questions for the moment while we consider some consequences. A serious potential difficulty is also illustrated by figure 1. If we consider center 1, 5 transfusions and no mismatches, the estimated survival is 103 per cent. Thus the model cannot be quite correct unless the effects are sufficiently small. Moreover, the effects cannot apply equally to all time periods. Even with small effects we would have to have different models for substantially different time periods. We will see that these objections can be met and still retain the advantages of an additive description.

The great advantage of additive models is that they allow
the description with a minimal and manageable number of
characteristics that must be determined from the data. At the
other extreme, the effect for each level of each factor would
have to be determined as a function of the levels of all of the
other factors, an impossible task if the relations are non-
trivial. Even in this case, the additive models would be
resorted to as providing useful, understandable summary
information. So we should pursue further the additive models.

In order to assure that the survival probabilities will
always fall in the range zero to one, it suffices to "stretch"
the scale of survival while retaining the additive expression.
Two transformations of scale are the more common, although the
possibilities are limitless. The "logit" scale, illustrated in
figure 2, expands the scale symmetrically about the 50 per cent
point. The additive portion of the model, illustrated in figure
3, proceeds as before to obtain a score, w; the per cent survival
is then calculated as $S=1/(1+\exp(-w))$. The result for the case
illustrated in figure 1, transplant center 10, 5 transfusions, 1
mismatch, is w=.32, S=58 per cent. The more extreme case, which
gave the impossible value of 103 per cent for the linear scale
additive model now gives 93 per cent. If we let the constant
depend upon time post transplant, the same set of effects could
apply to the entire post-transplant period. This would provide a
great economy in description. The general effect of the logit
scale is to shrink the linear scale difference as one approaches

the extremes of 100% and 0%, keeping a constant effect on the logit scale, as illustrated in figure 4. In the middle range, i.e. about 50%, there is little difference between the linear and logit scales; the increment on the per cent scale is about 25 times the increment on the logit - w - scale. The multiplier decreases at the upper part of the scale, so that a factor that increased survival from 35% to 65% for some set of conditions would result in an increase from 75% to 85% for an appropriate set of other conditions. The re-scaling accounts for the apparent decrease in effects of some factors as the level of survival increases, and does not necessitate the proliferation of magnitudes to describe the same effect operating at different levels of graft survival. We leave untouched the question of whether the constancy of effects on some scale or other reflects a deeper underlying biological significance. The logit scale has not been much used for the analysis of registry data, although Mickey(1,2) reported studies of cadaver donor transplants carried out on this basis. There is an extensive statistical development that is available for estimating the factor effects since the methodology is that associated with the "log-linear" models (3).

A much more widely used rescaling is the Gompertz scale. The mathematical relations are

$$w = \log(-\log S)$$

$$S = \exp(-\exp(w)).$$

The stretching of the linear scale is illustrated in figure 5. The comparison of this figure with figure 2, logistic scaling, emphasizes the asymmetrical nature of the Gompertz scale. The logistic and Gompertz are essentially the same for high values of survival but are greatly different for the low values. Figures 6-8 show the comparison of cadaver and parent donor transplants plotted on linear, logistic and Gompertz scales. If we are to speak of a constant effect of parent vs. cadaver donor, the curve for cadaver donors should be translated downward from that for the parent donor grafts. This, of course, is not the case for the linear scale — as it could not be and have both curves start at the same point at time 0. The curves are not parallel for the first two months for either the logit or the Gompertz scale, although after two months the magnitude of the difference is reasonably constant for the rest of the time shown. The comparison is not particularly discriminating since the two scales are quite similar over the range of survival plotted.

The three scales considered are special cases of additive models. A general formulation is given as

$$S(t|x) = G(\varphi(w(x)) + a(t))$$
$$w(x) = \beta_1 u_1(x_1) + \beta_2 u_2(x_2) + \dots + \beta_p u_p(x_p)$$

Logistic $\quad G = \dfrac{e^y}{e^y + 1} \quad y = \text{Log} \dfrac{G}{1-G}$

Gompertz $\quad G = e^{-e^y} \quad y = \text{Log}(-\text{Log}G)$

It is required that G be a non-increasing function and a(t) be a non-decreasing function. The structure of w(x) is shown as a regression form — i.e. it is an additive expression with terms given as products of regression coefficients — the β_i — and

scorings of the factors, the $u_1(x_1)$. There does not seem much to
be gained in pursuing the generality at this stage. Additivity
is sufficiently desireable a property, that it would seem more
useful to explore the possibilities. of other scales than to
abandon it.

The model most often used at present for formal
multivariable analysis (5,6,7,8,9,10,11,12,13,14) is the "Cox
regression" proportional hazards model (15,16). We have already
discussed this model, since the idea of the Cox regression is,
formally at any rate, the idea of re-scaling to the Gompertz
scale.

$$w = \beta_1 u_1(x_1) + \beta_2 u_2(x_2)$$
$$+ \ldots + \beta_m u_m(x_m)$$
$$S(t) = e^{-e^w H(t)}$$
$$Log(-Log\ S(t)) = w + Log\ H(t)$$

MODEL IS ADDITIVE ON GOMPERTZ SCALE

There is a well developed computation for estimating the unknown
constants — the β_i — for the Cox regression which is widely
available. An example result is given in table 1, a portion of
the result of an analysis of 5483 cadaver donor grafts
transplanted 1983-84 and reported to the UCLA Transplant
Registry. Most of us find such a presentation of results
somewhat unsatisfying. The values of the regression coefficient
estimates are not immediately transcribable into the more
familiar survival fractions. One tends to look at the ratios of

the estimates to their standard errors and in this way assess the statistical significance of the factors. We will reconsider the ideas of the Cox regression model in terms that are more readily assimilable than from table 1.

We may take risk, otherwise known as hazard, as the underlying idea. Risk is defined as the rate of failure. One could estimate the value as the number of failures in a (short) interval divided by the number of transplants that were at risk at the start of the interval and also divided by the length of the interval. The dimension of risk is failures per unit of time. The risk associated with the time marking the start of the interval is to be thought of as the limiting value of the ratio as the interval length shrinks to zero. The numerical relations are shown in the following equations.

RATE OF FAILURE. FAILURES/UNIT TIME/TRANSPLANT

$$h(t) \doteq \frac{S(t) - S(t+\Delta t)}{\Delta t S(t)}$$

$$= -\frac{d}{dt} \, \text{Log} \; S(t)$$

$$h(t) \doteq \frac{1}{\Delta t} \, \text{Log} \, \frac{S(t-\frac{\Delta t}{2})}{S(t+\frac{\Delta t}{2})}$$

The relation of risk to survival is illustrated in figure 9. In the case of kidney transplants, the risk drops off very rapidly in the early post-transplant period, then more slowly and after some two and half to three years becomes essentially constant. The risk function and the survival function are alternative expressions of the same thing, since each can be converted into the other.   The   survival curve is the more directly observable, however and is by far the more commonly used representation.

The idea of proportional hazards then is that the ratio of the risks corresponding to two sets of values of conditioning variables is a constant throughout the post transplant period

$$\frac{h(t|x_1)}{h(t|x_2)} = f(x_1, x_2)$$

$$h(t|x)\ \ = v(x)h_0(t).$$

If we denote the area under the risk curve as $H_0(t)$, then the area under the $h(t|x)$ curve is $v(x)H_0(t)$ and by taking logarithms of both sides we obtain

$$\log H(t|x)\ \ = \log v(x) + \log H_0(t).$$

Since H is equal to $-\log$ S, we are back to the model of additive on the Gompertz scale.   Thus we have shown that the Cox regression model of proportional hazards is essentially the same as the Gompertz scale additive model.

The ratio of estimated risks for parent and cadaver donor transplants, shown in figure 10, does not appear to be very constant.   There is considerable statistical variation in the estimates of risk, but even discounting this, the ratio appears to be larger in the early post-transplant period than later on. Since most of the failures occur early, the description in terms of constant risk ratio is probably useful regardless of whether the model strictly applies. The impression from figure 10 that the risk is fairly constant over the early post-transplant period is characteristic of a number of cases that we have plotted.

More to the point, perhaps, is that the average risk can be estimated fairly easily. The average risk from time t_1 to time t_2 is given as

$$\text{average risk} = \frac{\log S(t_1)/S(t_2)}{(t_2 - t_1)}$$

the log of the ratio of the survival fractions divided by the length of the time interval. In particular the average risk for the first year can be estimated as $\log 1/S$, where S is the one year survival fraction. Under the proportional hazards model the ratio of the average risks is the same as the ratio of the risks, so that one can estimate the risk ratio, the relative risk, directly. The results in table 2 illustrate the calculation and show that the numerical value of the log of the relative risk is reasonably close to the value obtained from a Cox regression analysis. Strictly speaking, the risk corresponding to the average of survival curves is not the average of the risks. The numerical results indicate that no serious error is involved in neglecting the distinction.

In identifying the average risk calculated from a grouping of transplants – e.g. parent donor grafts – we have ignored the multivariate nature of the problem and have relied upon an averaging process that would not be strictly correct. This seems a small matter as far as relating the somewhat abstract results obtained from the more detailed analysis to values one can readily calculate from the usual survival curves. Given a value for a base one year survival and numerical values for the Cox regression coefficients, one can obtain 1 year survival values by doing the calculations in reverse order. Thus the somewhat abstract results are not really so different from what one is accustomed to. The advantage of risk ratios is that they may be characteristic of the factor of interest. For example the ratio might be essentially the same for centers with relatively high success rates as for centers with low survival results, and in this way the ratio provides a compact, quantitative description of the comparison.

In describing the additive models formally, we have written the scoring of related factors as $u(x)$ corresponding to the value of the conditioning variable. This scoring is usually done as the relation $u(x)=x$, i.e. no rescoring at all. The scoring can make a great difference in the results, however, as is illustrated in figure 11. We show here the scorings $u(x)=x$, $u(x)=\log(1+x)$ and the empirically derived scoring developed (2) for the case of number of pre-transplant transfusions. Also shown is the 1-year average risk. The correlation with the first, $u(x)=x$, is quite weak as assessed by the square of the correlation coefficient, $R^2=.12$. The log transform is much better, $R^2=.59$, and the empirically based scoring is much superior, $R^2=.91$. The last is perhaps inflated since the correlations were computed using the same data set from which the empirical scores were derived. There can be considerable difference in the results depending upon the scoring used, and one of the tasks of analysis is to develop appropriate scores.

All of the models considered have been strictly additive. But this may not correspond to the reality. The effects of one factor may differ according to the level of other factors. Indeed much of the research is devoted to locating these "interactions". The essential character of the additive models can be retained by including terms that depend on pairs of factors only. Triplets or quadruplicates could also be included if needed, although the number of terms required would increase greatly if one attempted to add these systematically. Simplicity can be further obtained by forming the combined scoring as

$$u_{ij}(x_i, x_j) = [u_i(x_i) - \bar{u}_i] \, [u_j(x_j) - \bar{u}_j],$$

where the bars indicate averages. Alternatively, one might need to determine the relation on an empirical basis. There is good mathematical justification for using the product form if the joint effects are fairly small. Table 3 illustrates the results of using the product form for some 1983-1984 cadaver donor transplants. Of the interactions, only that for cyclosporine with regrafting appeared statistically significant. One year survival for the two-way table was estimated from the regression coefficients for cyclosporine, regraft and cyclosporine times regraft. The results are given in table 4 which also shows the observed survival for these groups of transplants. The risk ratio for cyclosporine is log 1/.723 divided by log 1/.654, (.324/.425)=.762 for first grafts; the log of the risk ratio is -.271. In order to calculate this from the regression coefficients of table 3, one needs to know further that cyclosporine was scored as .582 (the fraction of no cyclosporine) if cylosporine was used and −.418 (the fraction with cyclosporine) otherwise. These values were obtained by starting from the coding 1 and 0 and subtracting the average − the fraction with cyclosporine. Similarly the coding for regraft was .816 for regrafts and −.184 for first grafts. The contribution to the overall score w for cyclosporine, first graft is

$$w=(.364)(−.184) + (−.208)(.582) + (.333)(−.184)(.582)$$
$$= −.224$$

and the contribution for no cyclosporine, first graft is

$$w=(.364)(−.184) + (−.208)(−.418) + (.333)(−.184)(−.418)$$
$$= .046.$$

The log relative risk is then the difference (−.224)−(.046)=−.270, which agrees to within rounding error with the log risk ratio from the fitted values. Since the codings for both cyclosporine and regrafts are essentially on a 0 − 1 basis, the regression coefficient for the product term of table 3 is the difference of the log relative risk, cyclosporine vs no cyclosporine, between first and regrafts; it is also the difference between first and regrafts when conditioned on cyclosporine use. Accordingly, the regression coefficient for the product terms provides a statistical test for the uniformity of a factor conditioned on levels of the other factor entering the product.

The additive models may also be inappropriate because the risk ratio is not constant over time post-transplant, i.e. the proportional hazards model is incorrect. In terms of our development, the $u_1(x_1)$ are functions of time — $u_1(x_1,t)$. Provision is made for this possibility in the general computing procedures, but the implementation is open to the ingenuity of the analyst.

Once the proportionality of the risk functions is dropped, different approaches based on modeling of the risk functions appear to be interesting. One of the early analyses along these lines was that of Bailey and Homer (17), and also Krakauer et al. (18), in which the risk was expressed as the sum of a constant component and an exponentially decreasing component. The risk was described in terms of parameters that varied with the conditioning variables. Barnes and Olivier (19) considered a modified version. In this analysis the time post transplant was divided into intervals (0-3,3-12, and 12-24 months) and the assumption made that the risk was constant in each interval. The analysis was then concerned with estimating the contribution to the risk attributable to each of the several conditioning variables, in each of the periods. Gilks and co-workers (5,6) also divided the post-transplant period into intervals, (0-15, 16-40, 41-100, 100+ days) and considered the possibility of differing effects at the different periods. In addition they allowed for other possibilities of non-proportional hazards by carrying out a stratified analysis in which the basic risk was allowed to be different in selected subgroups — in their case, transplant centers.

It is probably a good idea to separate the post transplant period into intervals, as is indicated in the stylized plot of the risk function, figure 12. We indicate break points at about 4 months, 1 year and 2 years. The risk appears to decrease exponentially in the first period, as is indicated in figure 13 in which the risks in the early period are essentially linear and parallel when plotted on the log scale. The suggestion is that in this early period the risks are proportional and that the rate of decay of the risk may be uniform for a variety of subsets of the data. The period from about 4 months to a year is a transitional period during which the risks are no longer proportional. The transition period continues until about 24 to 30 months,after which the risks are essentially constant. The idea is quite similar to that of Bailey and Homer (17). Figure 13 alone does not establish these sweeping statements, but various other analyses we have done are in general agreement.

A more ambitious approach to statistical modeling is to consider a transplant as a random process in which the risk of failure at any given time is modeled to be dependent on the past history of the transplant – treatment and response to treatment, as well as values of variables determined pre–transplant. The development of Smith and West (20) for the problem of monitoring renal transplants is a step in this direction.

We have been concerned here with the ideas underlying the various statistical models for analyzing transplant registry type data and have not dealt with how to carry them out. For each of the additive models there are well developed and available computer programs, BMDP (21), SAS (22,23), for example, to carry out the analyses. For example, the linear scale additive models can be analyzed using standard linear regression programs, such as the BMDP 1R and 2R programs (20, pp 237-263). The logistic scale additive models can be analyzed using programs developed for log-linear models of categorical data, eg.BMDP4F (20, pp 143-206), and logistic regression analysis, eg BMDP LR (20, pp 330-344). In our use of the logistic additive model (2), we made use of a program based on the approach of Ruth Mickey (24). The proportional hazards models can be analyzed using programs such as BMDP 2L (20, pp 576-594). This program will also accommodate various possibilities of non-proportional hazards as well. The flexibility of the program is well illustrated by the analysis of Gilks et al. (5,6).

As we have presented them, the linear and logistic scale analyses apply to a fixed time, e.g. 1 year, and the programs mentioned would not take into account the actuarial aspects of data collection. One could construct a program that would actuarially accommodate the general additive model along the lines followed in the computation of the proportional hazards models, although there may not be enough of a demand to justify the development.

The difficult question seems to be that of the role of statistical models in transplant research. "What has been discovered from model based analysis that wasn't already known from other work?" "What, thought to be known from other work, has been shown to be incorrect on the basis of model based analysis?" If we had truly impressive achievements to point to, the questions would not be raised. But we are not entirely without answers. One distinctive feature of the model based approach is that since it is based on the idea of uniqueness of each transplant, it provides predictions of outcome for potential transplants (1,17). The possibility of predictions has not been vigorously pursued primarily because the predictions are not very discriminating. Distinguishing between 10% and 90% chance of 1 year survival would be useful. Distinguishing between 70% and 80% is of doubtful value. The question of why we cannot do better is a useful one but it is not clear how the question might be researched.

The most common answer to the question "Why use statistical models?" is that the models provide a basis for multivariable analysis. Since many factors influence transplant outcome, if only one of them is considered in isolation, what is the basis for confidence that the results are not the consequence of some combination of other factors that happened to be present in the material studied? There does not seem to be any way of answering the concern without carrying out the multivariable analysis. It appears to be a fact that the univariate analyses produce the correct results and that the concern is not well founded. This is particularly the case with the larger data sets where the effects of the other factors pretty well average out. One can give reasons why this would be the case; the correlations between the factors have to be fairly large in order for the masking effects to be pronounced. The concern may be appropriate for some of the smaller data sets.

Another answer might be that many of the findings currently available are of a qualitative/partly-quantitative nature. We ask: do transplants have better prospects if the recipient has been transfused prior to the operation? rather more frequently than we ask the quantitative question of the magnitude of the effect. In comparing results from several centers we are more interested in whether they all have the same general finding than whether the quantitative results are statistically/numerically consistent. The model based approach is more concerned with the quantitative aspects. We seek basic characteristics, such as risk ratios or logistic effects, that are descriptive generally. The question then becomes that of the desirability of quantitative approaches and raises questions of research strategy. If we seek quantitative information, the model based approach seems essential.

Finally one could say that any analysis is relative to some sort of a model. On this view one would argue that it is more fruitful to recognize the model that is being used and then seek analyses that will make efficient use of the available data. There is much to be said for this view, but it is also the case that mis-recognition of the underlying model might leave one worse off that he was before, or that undue concern with specifying the model might lead one to limit the scope of his thinking in regard to the depth of the study. These concerns are ways of stating that models aren't everything but that they have a prominent place in practically any analysis of transplant outcome data.

SUMMARY

Statistical models provide a framework for quantitative expression of the relation between outcome and conditioning variables. We noted the general desirability of additive models in that, to the extent that they are appropriate, they provide compact descriptions of the relation of outcome to individual variables that can be derived in the presence of the several variables under joint consideration. The model recognizes the uniqueness of each individual transplant. We illustrated the ideas with models that are additive on linear scale, on the logistic scale, and on the Gompertz scale, and gave expression to general additive models of which these three are special cases. We considered the concept of risk, or hazard, and showed that the proportional hazards model (Cox regression) is formally the Gompertz scale additive model. We showed that the risk averaged over a time interval can be easily estimated from grouped survival curve data. Although the grouping of transplants to form the survival curve is not strictly correct (since the risk function of the averaged survival is not the average of the risk functions for the grouped transplants) the results were shown empirically to be essentially the same. This leads to a more direct connection between the estimated regression coefficients in a proportional hazards analysis and the familiar survival curve. Models of the risk function were briefly discussed and models of the post transplant course viewed as a random process in which details of treatment and response are combined with pre-transplant variables to determine the unfolding risk were mentioned. Finally we briefly discussed the role of statistical models in transplantation research.

REFERENCES

1.   Mickey MR, Opelz G, Terasaki PI. Prospective estimates of success of kidney transplants: a basis for recipient selection. Transplant Proc 1979: 11:1914.

2.   Mickey MR, Multivariable analysis of one-year graft survival. In: Terasaki PI, Ed. Clinical Transplants 1985. Los Angeles, CA:UCLLA Tissue Typing Laboratory 1985:27-44.

3.   Bishop YMM, Feinberg SE, Holland PW. Discrete Multivariate Analysis: Theory and Practice. Cambridge: The MIT Press. 1975.

4.   Blagg CR. Treatment of end-stage renal disease by dialysis. Transplant proc 1985; 17:1497-1499.

5.   Gilks WR, Bradley BA, Gore WM, Selwood NH. Immunogenetic and clinical factors affecting renal transplantation. Transplantation 1986; 42:39-45.

6.   Gilks WR, Gore SM, Bradley BA. Analysisng transplant survival data. Transplantation 1986; 42:46-49.

7.   Krakauer H, Spees EK, Vaughn WK, et al. Assessment of prognostic factors and projection of outcomes in renal transplantation. Transplantation 1983; 36:372-378.

37

8. Kramer NC, Vaughn WK, Bollinger RR, et al. Comparison of cyclosporine and conventional immunosuppressive therapy in renal transplantation: a prospective multicenter study. Transplant Proc 1985; 17:2196-2201.

9. Rao KV, Umen AJ. Relative influence of risk variables associated with graft survival in cadaveric renal transplantation. Transplant Proc 1985; 17:2269-2272.

10. Sanfilippo F, Vaughn WK, LeFor WM, Spees EK. The relative influence of variables associated with cadaver renal transplant outcome. Transplant Proc 1985; 17:2256-2258.

11. Sanfilippo F, Vaughn WK, LeFor WM, Spees EK. Multivariate analysis of risk factors in cadaver donor kidney transplantation. Transplantation 1986; 42:28-34.

12. Sirchia G, Mercuriali F, Scalamogna M, et al. Factors influencing the cadaver kidney transplant program of North Italy. Transplant Proc 1985; 17:2259-2264.

13. Tiwari J, Mickey MR. Univariate and multivariate analysis of cadaver kidney graft survival data. in Terasaki PI, Ed. Clinical Transplantation 1986. Los Angeles, CA: UCLA Tissue Typing Laboratory 1986:231-246.

14.   Vereerstraetem P.   Magrez P,   Kinnaert P, et al. Comparison between actuarial and multivariate methods for analysis of kidney graft prognosis. Transplant Proc 1985; 17:2279-2282.

15.   Cox DR. Regression models and life-tables. J Roy Statist Soc B;34:187-220.

16.   Kalbfleisch J,   Prentice RL.   The   Statistical   Analysis of Failure Time Data. New York, Wiley, 1980.

17.   Bailey RC, Homer LD. Computations for a best match strategy for kidney transplantation. Transplantation; 23:329-336.

18.   Krakauer H,   Ed.   The   kidney   transplant histocompatibility study (KTHS):   analysis of   data: phase one: an   overview. Washington DC: ;US Printing Office, NIH publication no. 80-2164.

19.   Barnes BA, Olivier D.   Analysis of NIAID kidney transplant histocompatibility   study   (KTHS):   factors   associated   with transplant outcomes I. Transplant Proc 1981; 13:65-72.

20.   Smith AFM, West M. Monitoring renal   transplants: an application of the multiprocess Kalman filter. Biometrics 1983: 39:867-878.

21.   Dixon WJ Ed. BMDP Statistical Software. University of
California Press, Berkeley 1985.

22. SAS Institute Inc. SAS User's Guide: Statistics Version 5
Edition. Cary, NC: SAS Institute Inc., 1985.

23.   SAS Institute Inc. SUGI Supplemental Library User's Guide,
1983 Edition. Cary, NC: SAS Institute Inc., 1983.

24. Mickey RM, Elashoff RM. A Generalization of the Mantel-
Haenszel estimator of partial association for 2 X J X K tables.
Biometrics 1985; 41:623-635.

LEGENDS FOR FIGURES

Figure 1. Example illustrating model for per cent graft survival additive on linear scale.

Figure 2. Relation of logit scale to linear scale. Note that the "stretching" is symmetrical and extends indefinitely in both directions.

Figure 3. Example illustrating model for per cent graft survival additive on logistic scale.

Figure 4. Apparent shrinkage on the linear scale for an effect that is constant on the logistic scale.

Figure 5. Relation of Gompertz scale to linear scale.

Figure 6. Survival curves for parent donor and cadaver donor grafts plotted on linear scale.

Figure 7. Survival curves for parent donor and cadaver donor grafts plotted on the logistic scale.

Figure 8. Survival curves for parent donor and cadaver donor grafts plotted on the Gompertz scale.

Figure 9. Relation of risk function to survival curve.

Figure 10. Risk ratio, cadaver to parent donor, plotted against time after transplant.

Figure 11.   Contrast of different scorings of number of pre-transplant transfusions in relation to risk averaged over first year post transplant.

Figure 12.   Stylized risk function (based on cadaver donor transplants) with change points indicated.

Figure 13. Linearity and parallelism of risk curves during early post transplant period when plotted on plotted on log scale.

LEGENDS FOR TABLES

Table 1. Portion of result from proportional hazards analysis of cadaver donor grafts.

Table 2.   Estimation of relative risk from risk averaged over first   year   post-transplant   for .parent and cadaver donor transplants.

Table 3.   Results from proportional hazards analysis in which product (interaction) terms are included in the regression equation.

Table 4.   One year graft survival reconstructed from proportional hazards regression coefficients in comparison with fractions observed from grouped data for regraft vs cyclosporine.

Table 1. Portion of result from proportional hazards analysis of cadaver donor grafts.

FACTOR	$\hat{\beta}$	$\left|\dfrac{\hat{\beta}}{S.E.}\right|$
Graft No.	.364	5.17
Cyclosporine	−.208	3.32
Recip. Race	.167	3.13
ABDR MM	.058	2.81
Transfusion	−.130	2.26
...	...	...

Table 2. Estimation of relative risk from risk averaged over first year post-transplant for parent and cadaver donor transplants.

FACTOR	LEVEL	ONE YEAR GRAFT SURV. S	AVE. RISK FAIL./YR. LOG (1/S)
RELATION	CADAVER	66.2%	.412
	PARENT	87.5%	.134

RELATIVE RISK (RR) $= \dfrac{.134}{.412} = .324$ LOG RR $= -1.13 \pm .16$

REGRESSION COEFF., "COX REGRESSION" $= -1.05 \pm .14$

Table 3. Results from proportional hazards analysis in which product (interaction) terms are included in the regression equation.

FACTOR	$\hat{\beta}$	$\left|\dfrac{\hat{\beta}}{S.E.}\right|$
Graft No.	.364	5.17
Cyclosporine	−.208	3.32
Recip. Race	.167	3.13
ABDR MM	.058	2.81
Transfusion	−.130	2.26
GR × Cy	.333	2.95
GR × Race	−.200	1.57
GR × MM	−.052	1.40
Cy × Race	.218	1.57
Cy × MM	.055	1.63
Race× MM	−.045	1.22

Table 4. One year graft survival reconstructed from proportional hazards regression coefficients in comparison with fractions observed from grouped data for regraft vs cyclosporine.

(OBSERVED — N=5483,CAD,1983–84)

	CYCLOSPORINE	
	NO	YES
FIRST TRANSPLANT	65.4 (65.3)	72.3 (74.4)
REGRAFT	58.8 (57.3)	56.8 (58.6)

Figure 1. Example illustrating model for per cent graft survival additive on linear scale.

CONSTANT = 60

CENTER EFFECT		TRANSF. EFFECT		MISMATCH EFFECT	
1	25	0	−15	0	8
2	4	1	−1	1	2
...		...		...	
10	−15	5	10	6	−9
...		...			

CHANCE OF GRAFT SURVIVAL FOR 1 YEAR

60 − 15 + 10 + 2 = 57%

Figure 2. Relation of logit scale to linear scale. Note that the "stretching" is symmetrical and extends indefinitely in both directions.

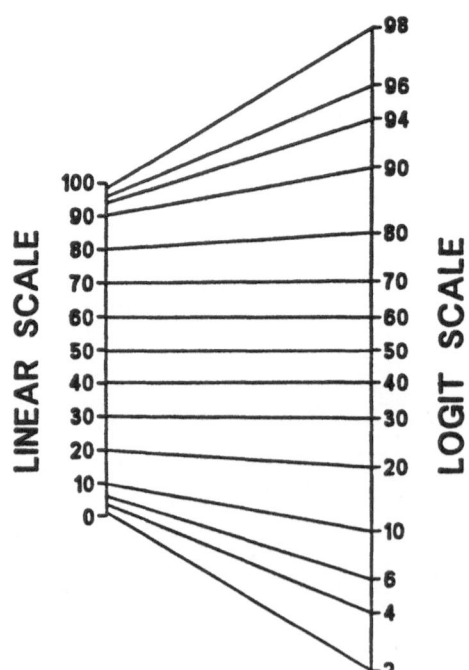

Figure 3.   Example illustrating model for per cent graft survival
additive on logistic scale.

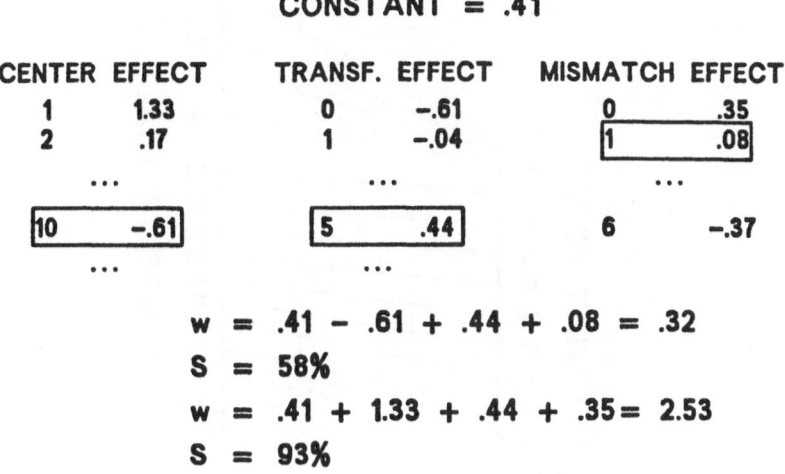

CONSTANT = .41

CENTER EFFECT		TRANSF. EFFECT		MISMATCH EFFECT	
1	1.33	0	−.61	0	.35
2	.17	1	−.04	1	.08
...		...		...	
10	−.61	5	.44	6	−.37
...		...			

w = .41 − .61 + .44 + .08 = .32
S = 58%
w = .41 + 1.33 + .44 + .35 = 2.53
S = 93%

Figure 4.   Apparent shrinkage on the linear scale for an effect
that is constant on the logistic scale.

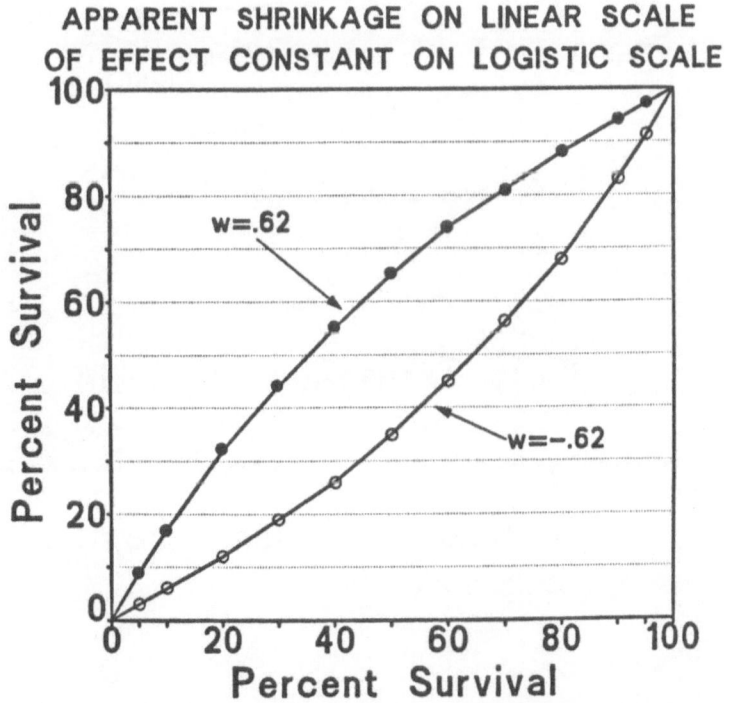

APPARENT SHRINKAGE ON LINEAR SCALE
OF EFFECT CONSTANT ON LOGISTIC SCALE

Figure 5.   Relation of Gompertz scale to linear scale.

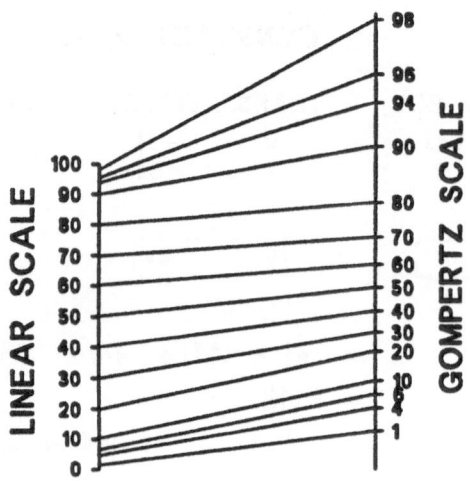

Figure 6.   Survival   curves for   parent donor   and cadaver donor grafts plotted on linear scale.

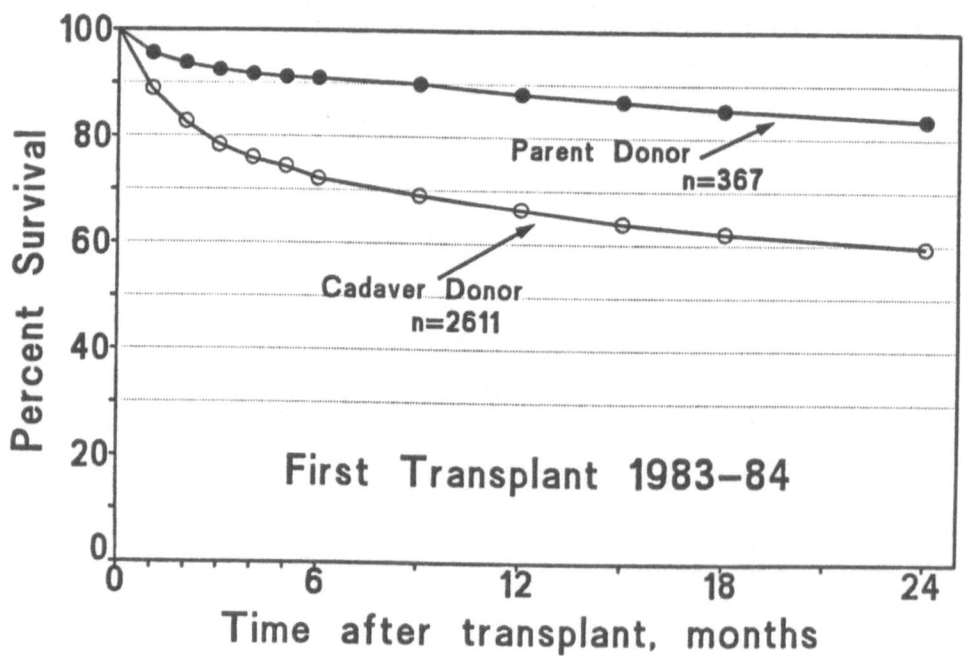

Figure 7.   Survival   curves for   parent donor   and cadaver donor grafts plotted on the logistic scale.

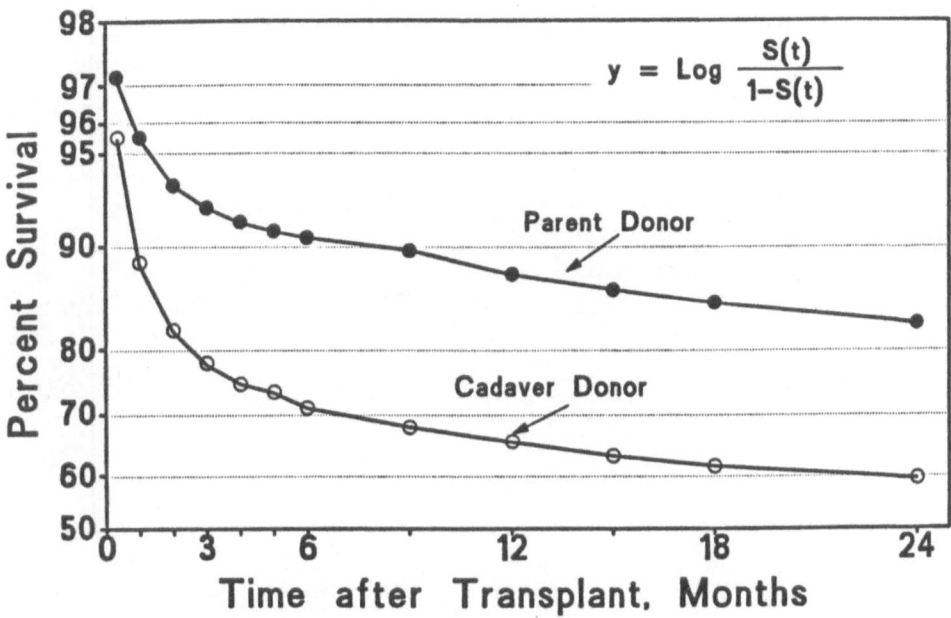

Figure 8.   Survival   curves for   parent donor   and cadaver donor grafts plotted on the Gompertz scale.

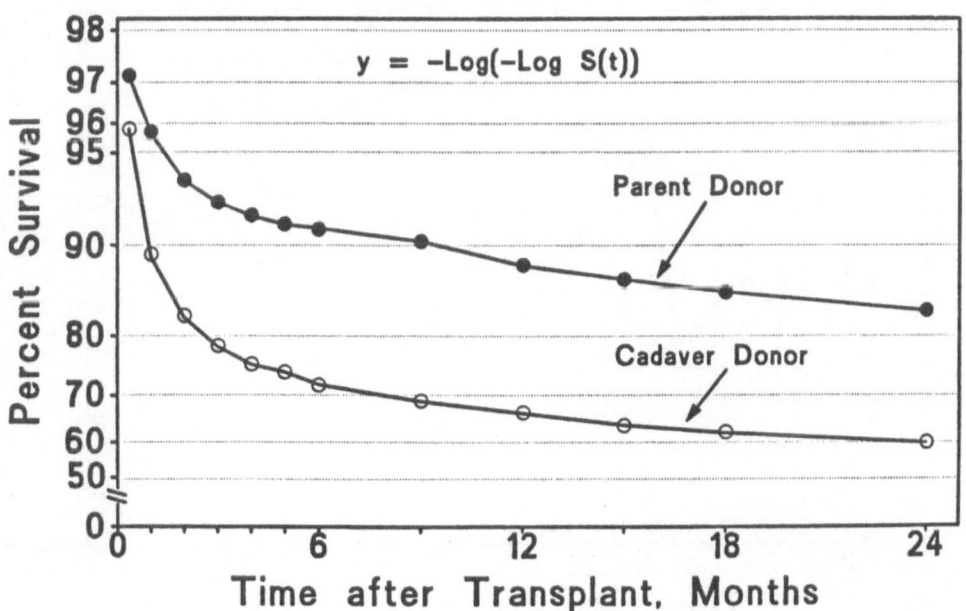

Figure 9.   Relation of risk function to survival curve.

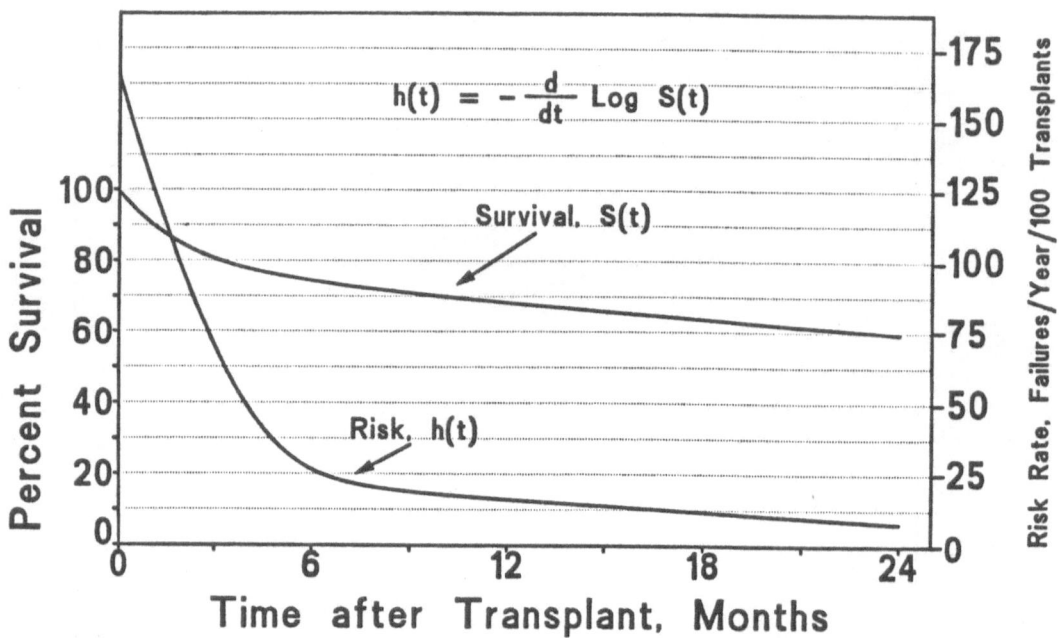

Figure 10.   Risk ratio, cadaver to parent donor, plotted against time after transplant.

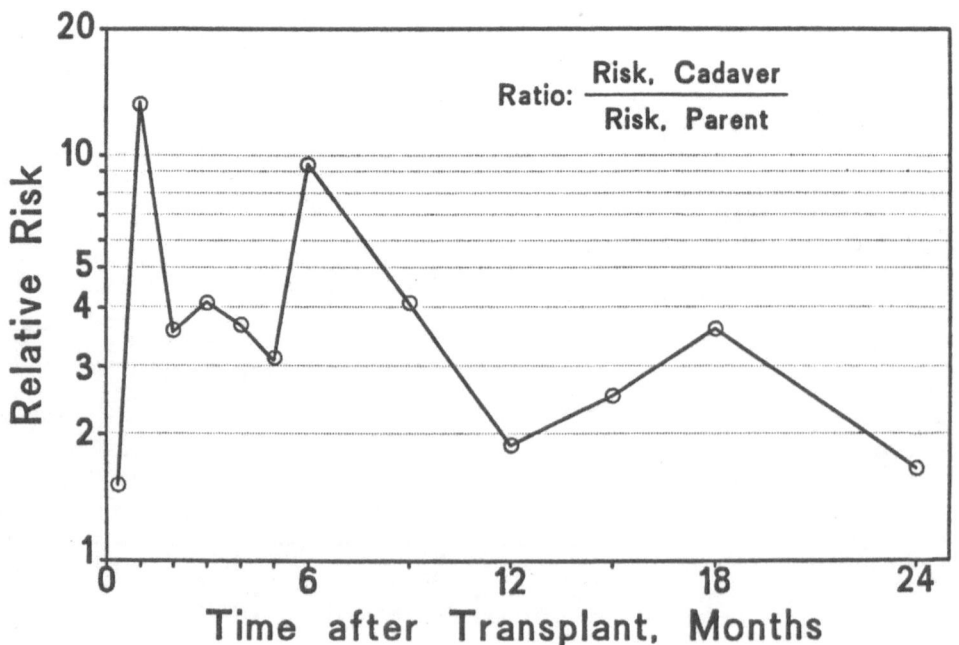

Figure 11. Contrast of different scorings of number of pre-transplant transfusions in relation to risk averaged over first year post transplant.

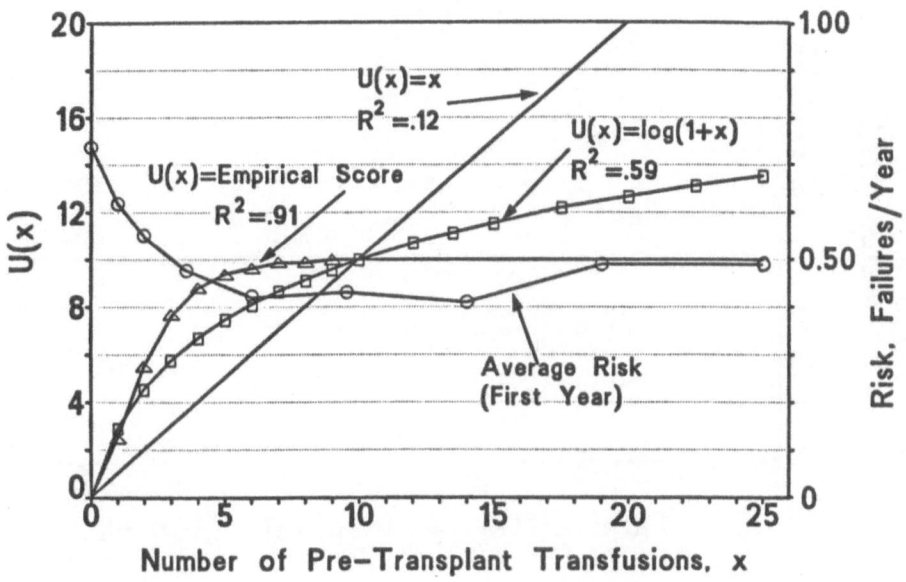

Figure 12. Stylized risk function (based on cadaver donor transplants) with change points indicated.

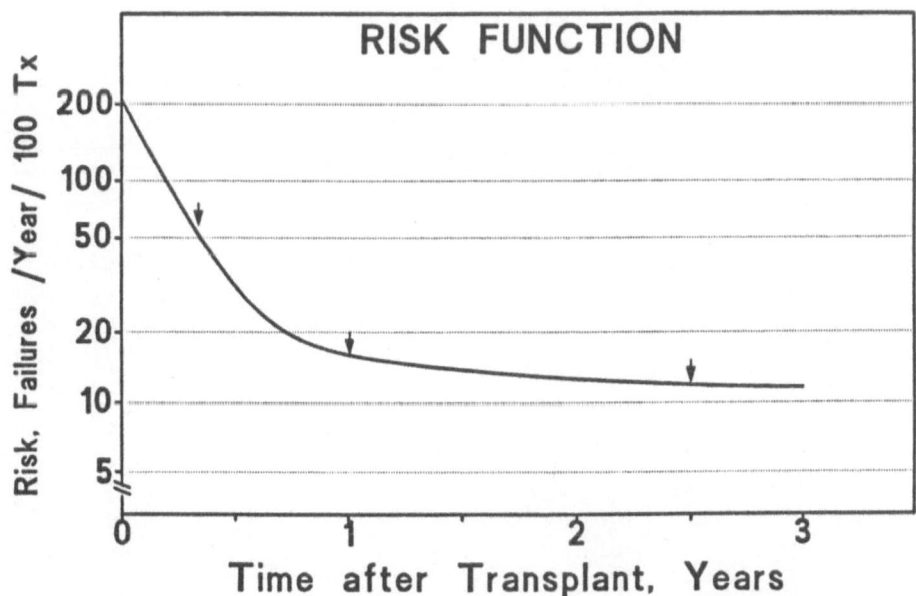

Figure 13.   Linearity and parallelism of risk curves during early
post transplant period when plotted on plotted on log scale.

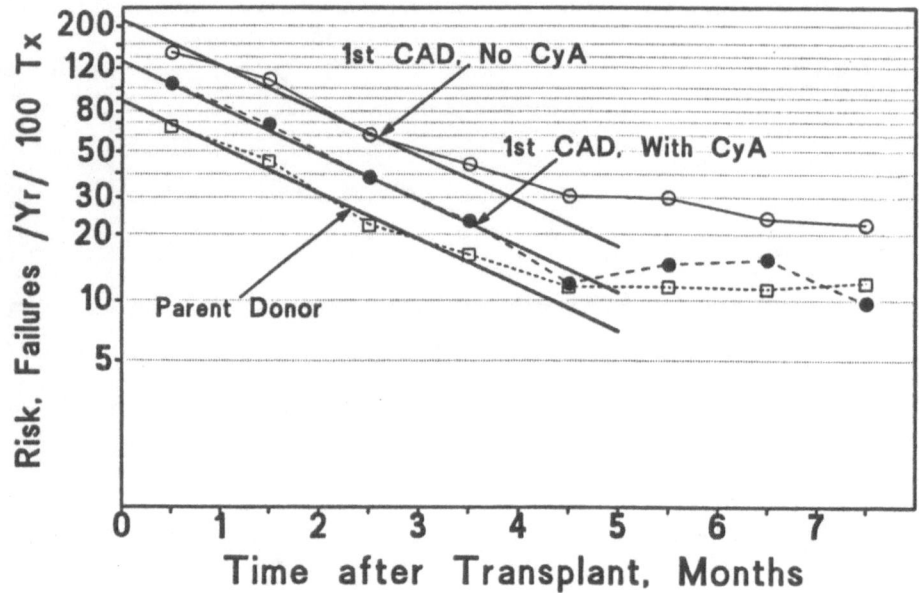

STRATEGIES FOR MULTIVARIATE DATA ANALYSIS: CASE STUDIES

Herman P. Friedman
IBM Corporate Technical Institutes
Systems Research Institute
500 Columbus Avenue
Thornwood, NY 10594
USA

ABSTRACT

Much of what we do in data analysis today is not formalized. We have been able to describe and implement in computer software many powerful technical tools. What is required is a way to formulate, describe, analyze and evaluate strategies for doing data analysis. A start in this direction is to describe case studies in such a way as to make explicit the rationale for the choices open to the data analyst.
In this spirit, this paper describes analyses of measurements taken on surgical patients. Tools of multivariate statistics, pattern recognition, and computer graphics were used for analysis and graphical representation. Results were evaluated in a medical context. Implications from these studies will be drawn with regard to the role of statistical data analysis in the development of Expert and Decision Support Systems.

L. J. Savage (1967) suggests: "Many of the most tantalizing things in statistical work today could be called descriptive statistics. These are efforts to arrange and condense complicated bodies of data in ways that promise you a fighting chance to see what is essential."

INTRODUCTION

There is a need for statisticians to develop broad guidelines for approaching data analysis in statistical problems. A start for such a framework was provided by D. Cox (1986). A. Dempster (1982) used the term "functional statistics", to mean the process

of integrating the technical tools of statistics with the subject
matter. He indicates that little of this is taught to the statistician. Much of what we do in data analysis today is not
formalized. While we have been able to describe, formalize and
implement in computer software many powerful technical tools, we
still require a way to formulate describe, analyze and evaluate
strategies for carrying out data analysis. A start in this direction is to describe case studies in such a way as to make explicit
the choices open to the Data Analyst.
In the past , it has been difficult to publish or present details
of case studies. Indeed the existence of choices at each stage of
investigation were rarely mentioned in print, for fear of starting
a debate. The recent interest in the application of methods of
artificial intelligence to problems of data analysis has provided
an impetus for describing the process of carrying out Data
Analysis.
In this spirit, this paper describes two interrelated case studies.

CASE STUDY 1

The objective of the first analysis was to develop a classification
of physiologic states in surgical patients.
A data bank of patient information was available from the surgical
service of a large metropolitan hospital for cases that were
studied and treated in a Clinical Research Center. These data
originated from a large number of sources, including bedside
cardiac catheterizations, patient records, laboratory determina-
tions, and physicians' and nurses' notes. These patients were
studied over a time period ranging from less than 1 day to nearly 1
month. During the course of treatment on an individual patient,
anywhere from 1 to 50 detailed sets of measurements were collected
as required for patient treatment and not according to a fixed
protocol. Each set of measurements, which is centered about a
determination of cardiac output, can involve up to 151 primary
variables of the type indicated above, and may also produce up to
61 additional derived variables. The computer-based data retrieval

system used in this study and examples of the multivariable data collected have been previously reported by Siegel et al., (1972). The data are cross-sectional with regard to the measurement sets of all patients and longitudinal with regard to many sets on an individual patient. Note that the various measurements at any time point for an individual patient are keyed to cardiac output. The time intervals between measurements on a patient are not equal, the number of measurement sets varies from patient to patient, and the physician's clinical description of the patient may change from one time period to another.

In collaboration with the physician and after some preliminary screening, a coherent set of eleven variables that contain information concerning both the state of the heart and of the peripheral circulation was selected. These variables were: cardiac index (CI), heart rate (hR), mean blood pressure (MBP), arteriovenous oxygen difference (A-V Diff), systolic ejection time (ET), right atrial mean pressure (CVP), mixed venous pH (VpH), mixed venous pO_2 (VpO_2), mixed venous pCO_2 ($VpCO_2$), and two parameters derived from the analysis of the shape characteristics of the central indicator dilution curve, the cardiac mixing time (tm), and the dispersive time (td).

The results of the data analysis produced a reference control group (R state) and a scaling that allows the entire spectrum of clinical severity in patients with trauma, sepsis, or cardiogenic shock to be viewed in terms of four pathophysiologic states (A, B, C, and D). The prototype patient in each of these abnormal states can be described by the physiologic pattern of the multivariate means (Figure 1). In this figure each pattern (dark line) is compared to the perfect circle (heavy circle) of the averages of the R state. Differences are in standard deviation units of the R state.

The following is an outline of the stages in the analysis that led to the desired classification:

(1) Define by clinical criteria a basal reference group.

(2) Robust estimation of scale and location parameters.

(3) Scale data with respect to reference group.

(4) Use minimum variance partition criteria (K means algorithm) to cluster states.

(5) Define prototype states R, A, B, C, D.

The key steps in this analysis were:

- Selection of the basal reference group;
- Development of prototype states for patient evaluation;
- Choosing of variables for analysis in collaboration with physician;
- Understanding the observational nature of the data recorded from patients during the hospital stay.

We define the data set for analysis:

$X(I, J, K)$...

$I = 1, 2, .., N$	Patients
$J = 1, 2, .., 11$	Variables
K = Index of Time	Unequal No. of Intervals;
	Unequal Interval Length for Each Patient

The identification of a basal reference group was based on clinical records with the initial choice made by the physician. That is, at a given point in time, the patient was said, by the physician, to have no clinical abnormalities.

In mathematical notation, we have:

$(X(I, J, K^*))$ are those measurements for which the patient is defined "Normal" by the physician.

From this data set, we performed robust estimations of the mean vector and scale parameters.

There are many choices, both formal and informal, for the estimation of scale parameters and outlier detection. There are a variety of formal procedures for robust estimation of scale and location parameters (e.g. Seber (1984). These formal methods include the fitting of multivariate T distributions. Informal procedures include the use of histograms and scatter plots, probability plots of mahalanobis distance, and single linkage cluster analysis.

The determinant of the within-group covariance matrix as a cluster criteria is useful in detection outliers. K-means clustering is also useful in this way. By choosing K large, we found that outliers appear in small groups.

CASE STUDY 2

The main thrust of the analysis was to characterize patterns of coronary by-pass patients based on physiologic measurements taken serially on patient preoperatively, during the operation and postoperatively on the intensive care unit. Questions of interest are:

Is there a typical or "normal" recovery pattern for coronary by-pass patients?

Are there identifiable patterns of departures from this "normal" recovery pattern?

If yes, can these patterns be meaningfully related to outcome measures and modes of treatment?

The primary physiologic measurements on an individual patient are all measured at the time the cardiac output is determined.

The time between measurement sets (interventions) is not uniform within an individual patient.

The times between individual patients do not correspond directly.

Again many of the same issues described in the prevous case study have to be considered by the Data Analyst. The purpose is to characterize "typical" recovery pattern of patterns of coronary by-pass patients. The Data Structure is described by:

$M (I, J, K)$, $I = 1, ..., N$, for patients

$J = 1, ..., V$, for variables

$K = 1, ..., L (I)$, K is a time index.

The number of intervals for each patient may be different. The
time points for each patient are not all comparable as in the
previous study.

The overall strategy was to select a frame of reference that would
allow for a meaningful comparison of patients cross-sectionally at
selected time points and yet would capture the longitudinal dynamic
response patterns.

The number of measurement sets (interventions) varied from patient
to patient. Those patients that did poorly will have the largest
number of interventions. However, all patients studied have had a
preoperative measurement set, at least one immediate postoperative
intervention (within a few hours after surgery), and at least one
intervention on the day following surgery.

A set of patients were chosen for analysis that had complete data
at four time points. The time points were preoperative, immediate
postoperative, the next day following surgery, and the last
measurements before the patient left the intensive care unit. The
data set analyzed is:

      DATA (I, J, K*),   I = 1, .., 72 patients
                      J = 1, .., 11 variables
                      K* = 1, 2, 3, 4, time

The key steps of the analysis were:

1) Selection of an initial set of measurements and a set of
   initial time points. (preop, imedpostop, next, last in ICU).
2) Histograms and Scatter Plots for exploratory data analysis. (Figures 2,
3) Discriminant analysis among preop, imed, last day in the ICU.
4) Projection onto discriminant plane obtained above. (Figure 6)
5) Characterization of trajectories in the plane by 8 coordinates
   (reduction in dimensionality from 44 to 8 measurements).
6) Robust cluster analysis of 8 dimensional data (trajectories).
7) Use of principal components to reduce 8 dimensions to 3.
8) Use of 3D Graphics to corroborate cluster analysis. (Figures 4,5)
9) Relate to original variables.
10) Graphical descriptions of typical trajectories. (Figure 7)
11) Interpretation and relation of recovery patterns to previously
    understood physiological states. (Figure 7)

In earlier stages of our analysis we used graphics for histograms and scatter plots to get a rough idea of location and scale parameters as well as to find a few "wild" data points. Anscombe (1973) describes a useful scatter plot that allows the user to define plotting symbols based on coded values of a third variable.

Later on in the analysis, graphs are needed as a natural adjunct of the analysis techniques. To cope with the high dimensionality (many variables), a discriminant analysis was performed among the preoperative, postoperative and the last day in the intensive care unit. A plot in the space of the canonical variables is shown in Figure 2. The plot is shown labeled with the symbols P, I, N, and L. Connecting the points for each patient led to a very messy, unreadable graph. However we had achieved a reduction of dimensionality from 44 to 8. Namely for each patient at each of four time points we have two discriminant function scores. The patient trajectories were then clustered using a K-means algorithm. After clustering, a number of plots were made to aid in interpreting the results. The clustering procedure produced an index vector in APL that made the rest of the analysis less labor intensive. The results of the various cluster analyses showed one or two "typical" patient trajectories with numerous very small clusters of "atypical" trajectories. To describe typical trajectories, lines were drawn in the space of canonical variables. Thus, cluster analysis transformed a messy set of lines into a few usable graphs. See Figures.

To deal with the presence of outliers, a principal component analysis of the 8-dimensional space was performed. Three dimensional plots of the first three components were made. Figures represent two different perspective views of the same data. Robust estimates of the first two components were obtained and a 3-D plot was made with the third coordinate a suitably scaled value of the residuals.

These 3-D plots were used to check on the "atypical" trajectories found by clustering. The plots provided visual evidence to support the findings of the cluster analysis.

Robust methods for multivariate analysis are not readily available
in computer packages. A program was written in APL for a recursive
method of fitting a multivariate T distribution to provide robust
estimates of the covariance matrices required. Robust methods
require examination of the data and judgment as to choice of
weights and/or models. Graphical methods are necessary for making
intelligent choices.

To gain further insight and understanding, graphs and analyses were
made in terms of the original variables. Plots over time for each
patient and prototypes were made using graphs of the type in Figure
1 and described in Siegel et al (1972).

SUMMARY OF DATA ANALYSES:

There are a number of points that we would like to emphasize.
A critical factor in the analysis of the data is to exploit some
feature of the data that can provide a frame of reference. In the
first case study we used the concept of a "basal reference state"
to provide a template to view the patients. The graphs shown in
Figure 1 are scaled polygons. Without the template provided by the
analysis, the graphs would be uniformative.

The analysis in case study two added credibility to the "Basal
Reference Group" (R) developed in study one. The patients in the
second case study were from a different hospital all undergoing
coronary by-pass surgery. We found that preoperatively the
patients could be compared to the D group, and the day they leave
the intensive care unit they look more like the A group (a compensated stress group), than like the reference group R. Similar
observations of the stress response to surgery were reported by
Shoemaker (1973), and by Siegel et al (1980).

IMPLICATIONS FOR THE DESIGN OF EXPERT SYSTEMS

A major issue in the design of decision support or expert systems
is the development of a knowledge base. This is loosely described
as eliciting from the expert (e.g., the physician) his or her
"knowledge". The strategies suggested by the two studies described
here as well as in a study of the Borderline Syndrome in
psychiatry, reported, by H. P. Friedman (1968), indicate that the
physician is exellent in identifying the relevant measurements and
in recognizing patterns of these measurements. What is required,
is the collaboration of computer scientists, statistical scientists, and the field expert to develop a reference classification
system with appropriate scaling that is reproducible and useful.
Other relevant approaches can be found in Sneath (1975) and
Thompson and Woodbury (1970). A publication by W. Gale (1986)
contains many articles discussing the relationship between statistics and artificial intelligence.
We now summarize the key elements in this activity and key decision
points that must be determined collaboratively.

    Purpose of Classification
    Type of Classification - Cross-Sectional of Longitudinal
    Selection of patients (Cases) or subjects
    Selection of variables (Issues of High Dimensionality)
    Scaling of Variables
    Choice of Clustering Criteria
    Choice of Clustering Algorithms
    Evaluation of Clusters
    Graphical Representation of results
    Interpretation and Relation of Clusters to Prognostic and
     Clinical Information.

REFERENCES

1. Anscombe, F. S., (1973, "Graphs in Statistical Analysis,"
    American Statistician, 2/27/73, pp. 17-21

2. Anscombe, F. A., (1981), Computing in Statistical Science
    Through APL, Spring-Verlag, 1981.

3. Cox, D. R., (1978), "Some Remarks on the Role in statistics of
    Graphical Methods," Applied Statistics, 27, No. 1, pp. 4-9,
    1978.

4. Cox, D.R. (1986) "Some General Aspects of the Theory of
    Statistics" International Statistical Review, 54, 2, pp
    117-126.

5. Dempster, A., "Purposes and Limitations of Data Analysis,
    "Research Report S-85, 1981, Department of Statistics, Harvard
    University.

6. Friedman, H.P. and J. Rubin (1967), "On Some Invariant Criteria
    for Grouping Data," J. of the American Statistical Association,
    Vol. 62, No. 320, pp. 1159-1178, December, 1967.

7. Friedman, H.P. and J Rubin (1968), "The Logic of the Statis-
    tical Methods" Chapter 5 of The Borderline Syndrome by
    R. Grinker Sr., B. Werble, R. Drye. Basic Books Inc.

7. Friedman, H. P., Goldwyn, R. M., Seigel, J. H. (1975), The use
    and interpretation of Multivariate Methods in the Classifica-
    tion of stages of Serious Infectious Disease Processes in the
    Critically Ill. In Blashoff, R. (ed.): Perspectives in
    Biometrics, New York, Academic Press, 1975, pp. 81-122.

8. Gale. W (1986) Artifical Intelligence and Statistics. Addison
    Wesley Press 1986

8. Savage, L. J. (1967), Conference on the Future of Statistics:
    Proceedings, edited by D. Watts, Academic Press, p. 146.

9. Seber, G. A. F. (1984), Multivariate Observations, John Wiley &
    Sons, 1984.

10. Shoemaker, W. C. et al (1973), "Physiologic Patterns in
    Surviving and Nonsurviving Shock Patients", Arch. Surg./Vol.
    106, May, 1973.

11. Seigel, J., et al (1972), "The Surgical Implication of
    Physiologic Patterns in Myocardial Infarction Shock." Surgery
    Vol. 72, No. 1., pp. 126-141, July, 1972.

12. Siegel, J., et al (1980), "The Effect on Survival of Critically
    Ill and Injured Patients of an ICU Teaching Service Organized
    About a Computer-Based Physiologic CARE System," The Journal of
    Trauma, Vol. 20. No. 7, 1980, pps. 558-579.

13. Sneath, P.H.A. (1975), "A vector Model of Disease for Teaching and Diagnosis. - <u>Medical Hypothesis</u>, March-April, 1975.

14. Thompson, H. K., Jr., Woodbury, M.A., "Clinical Data Representation in MultiDimensional Space." <u>Computers and Biomedical Research, 3, 58</u>, 1970.

Physiologic patterns representing prototype patients in septic and cardiogenic shock states

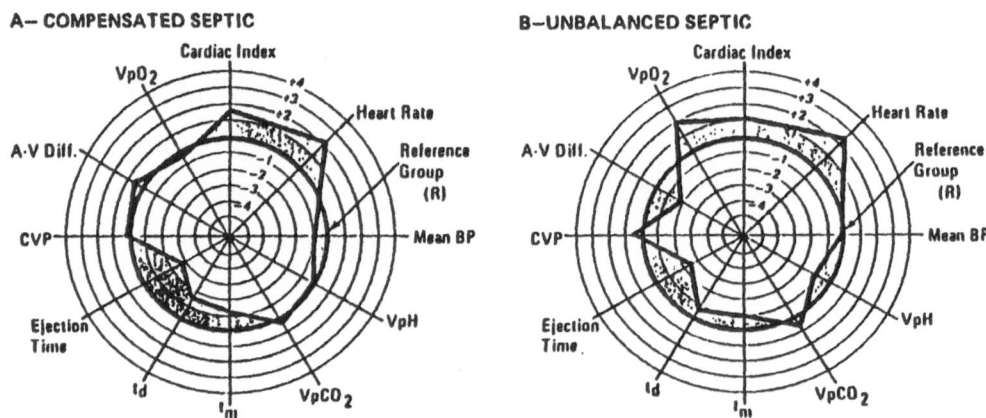

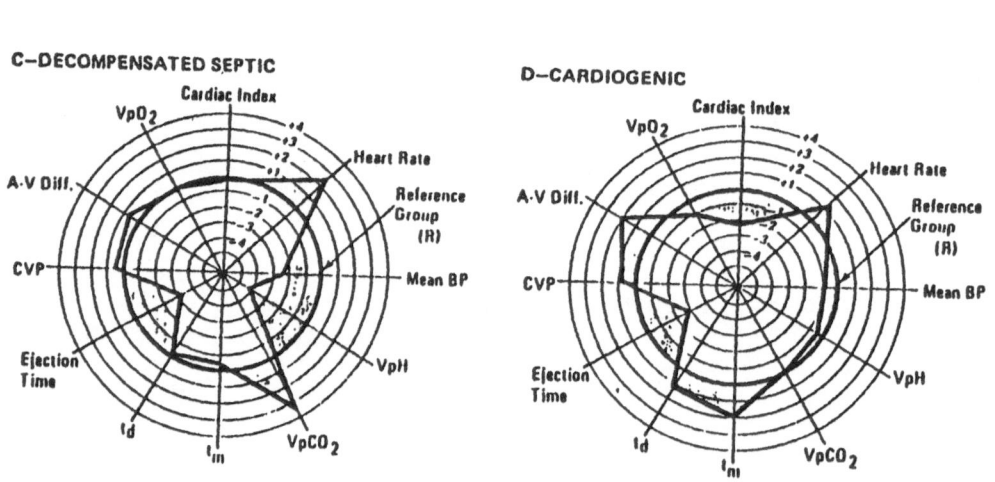

Figure 1

63

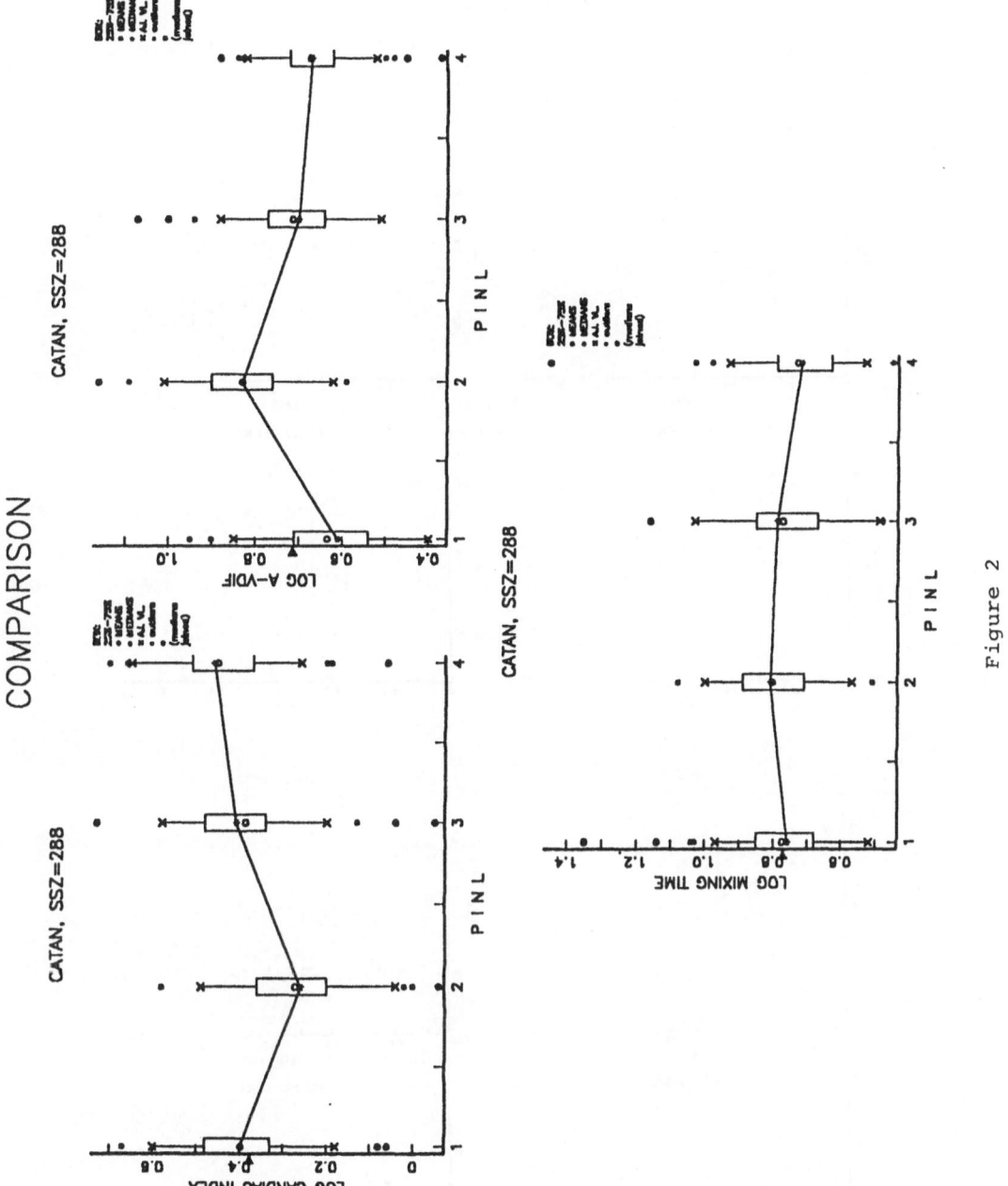

Figure 2

Figure 3

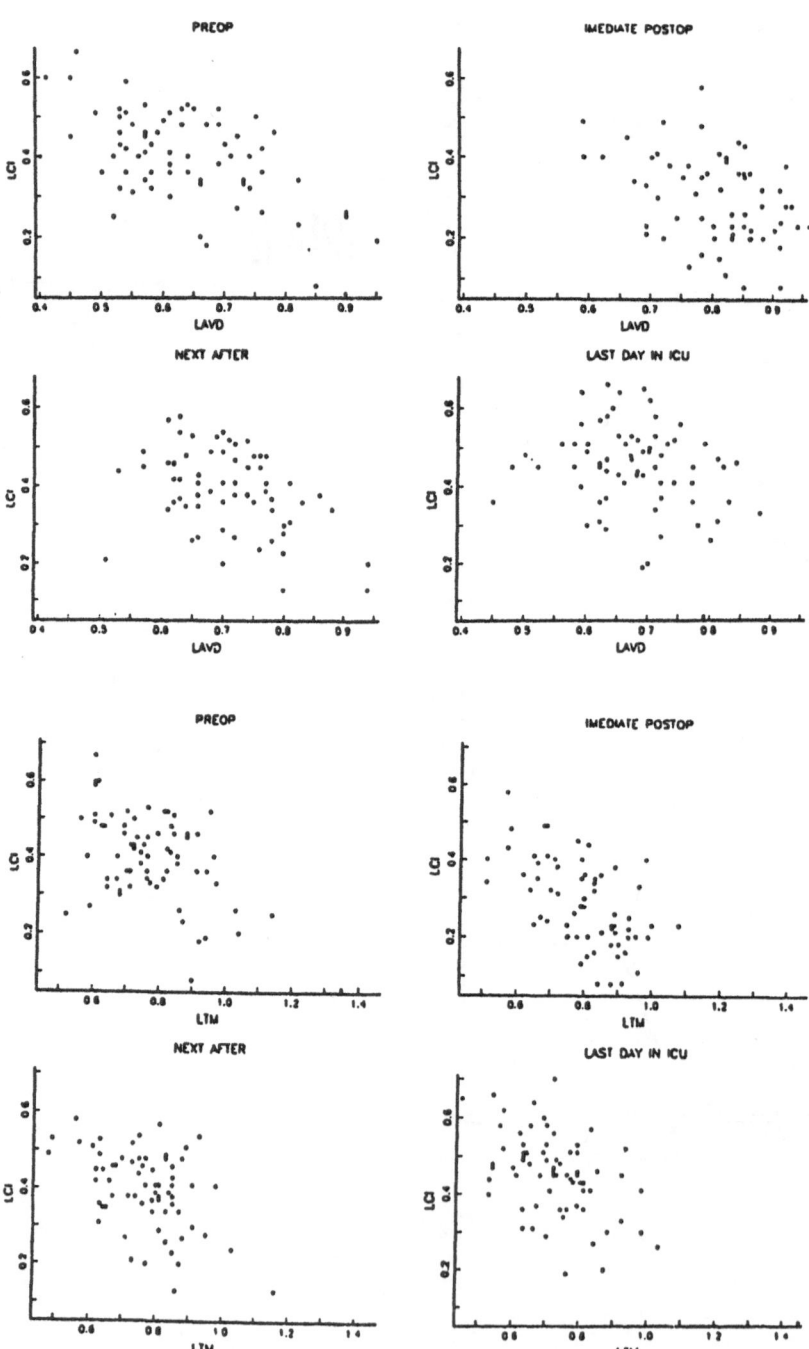

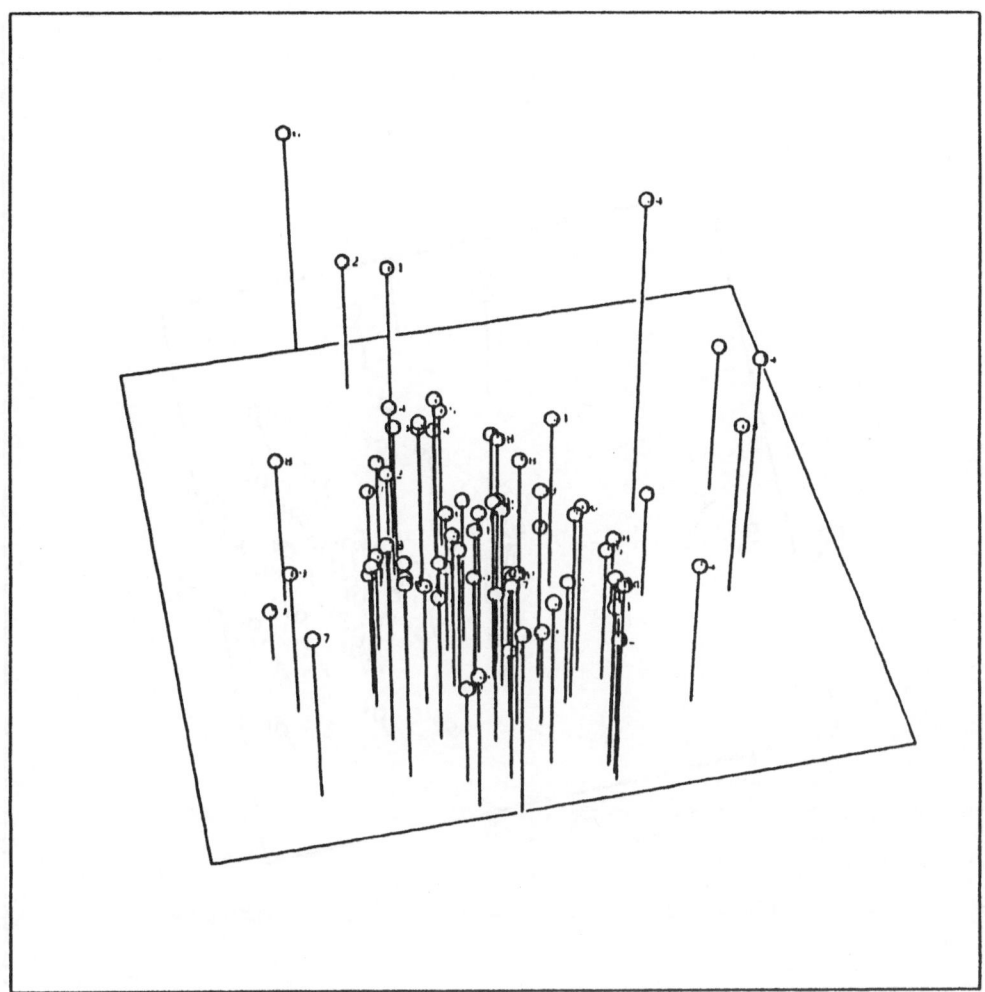

Figure 4

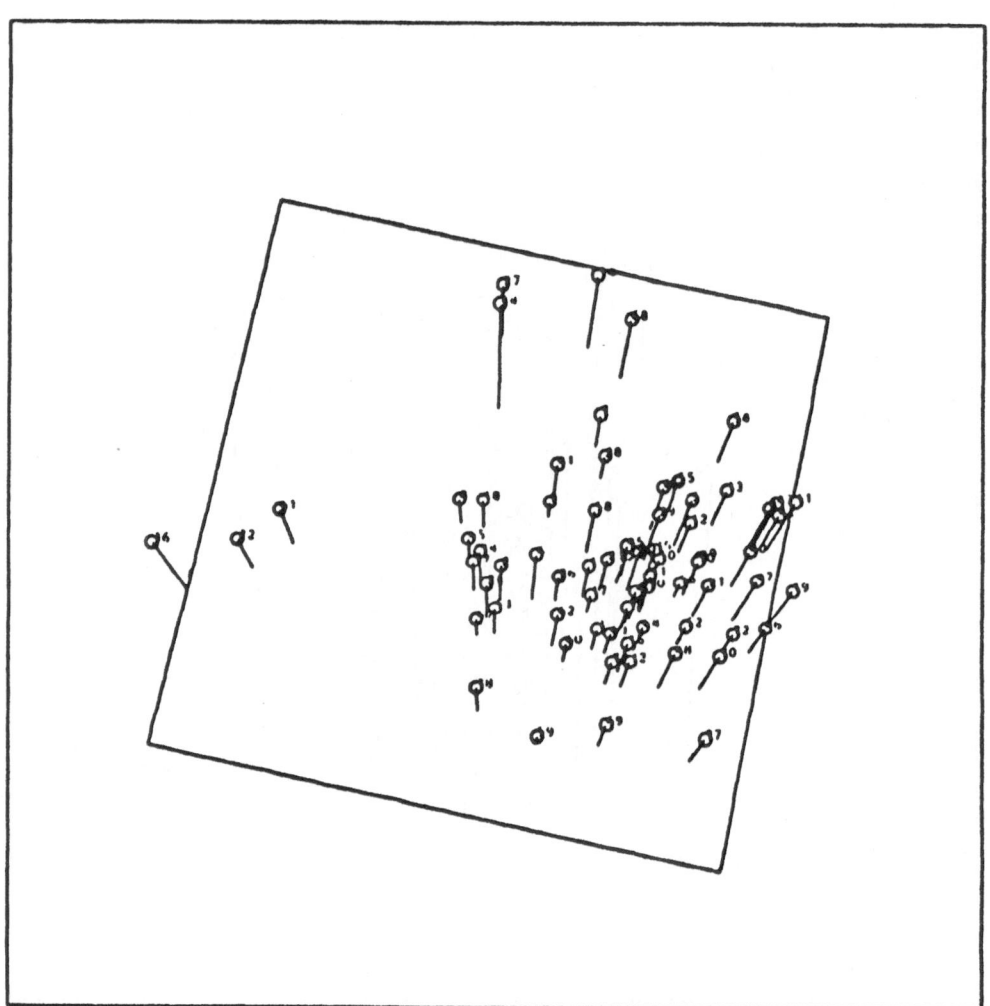

Figure 5

PROJECTION ONTO DISCRIMINANT PLANE OF PIL

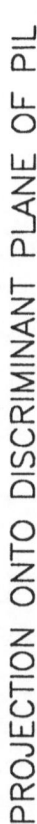

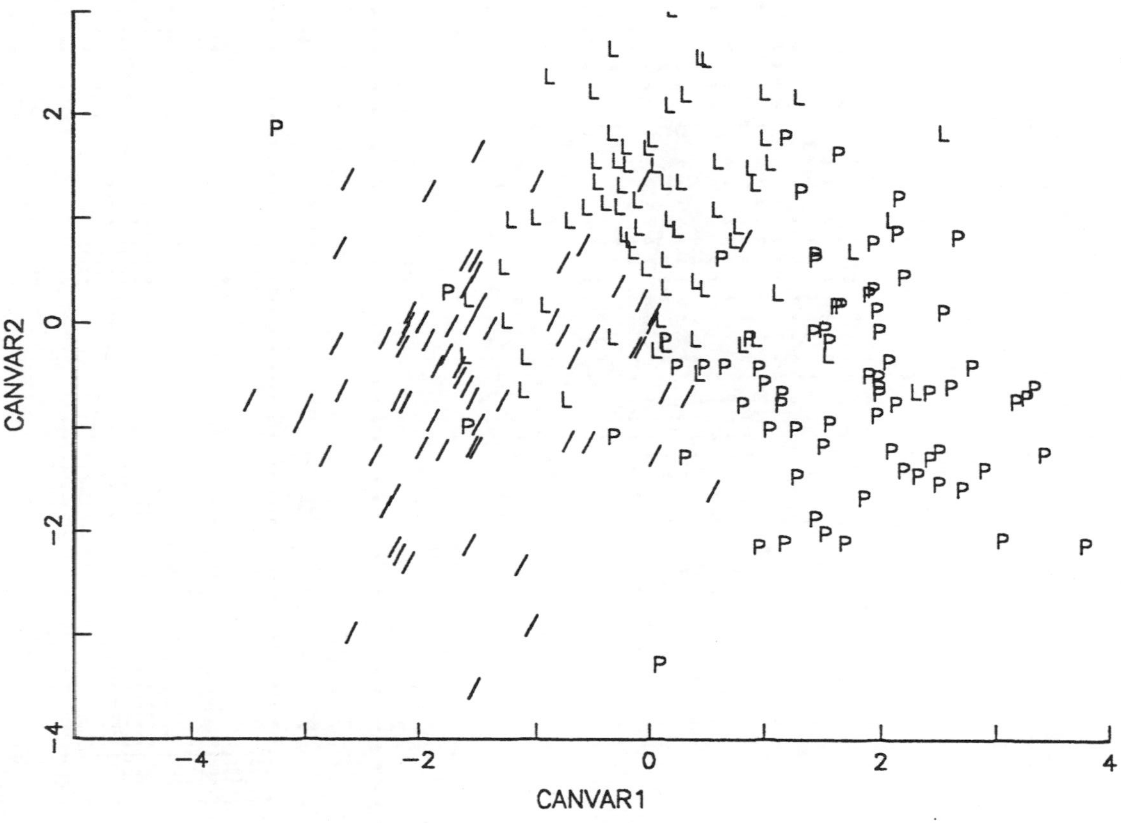

CANVAR1

CANVAR2

Figure 6

Figure 7

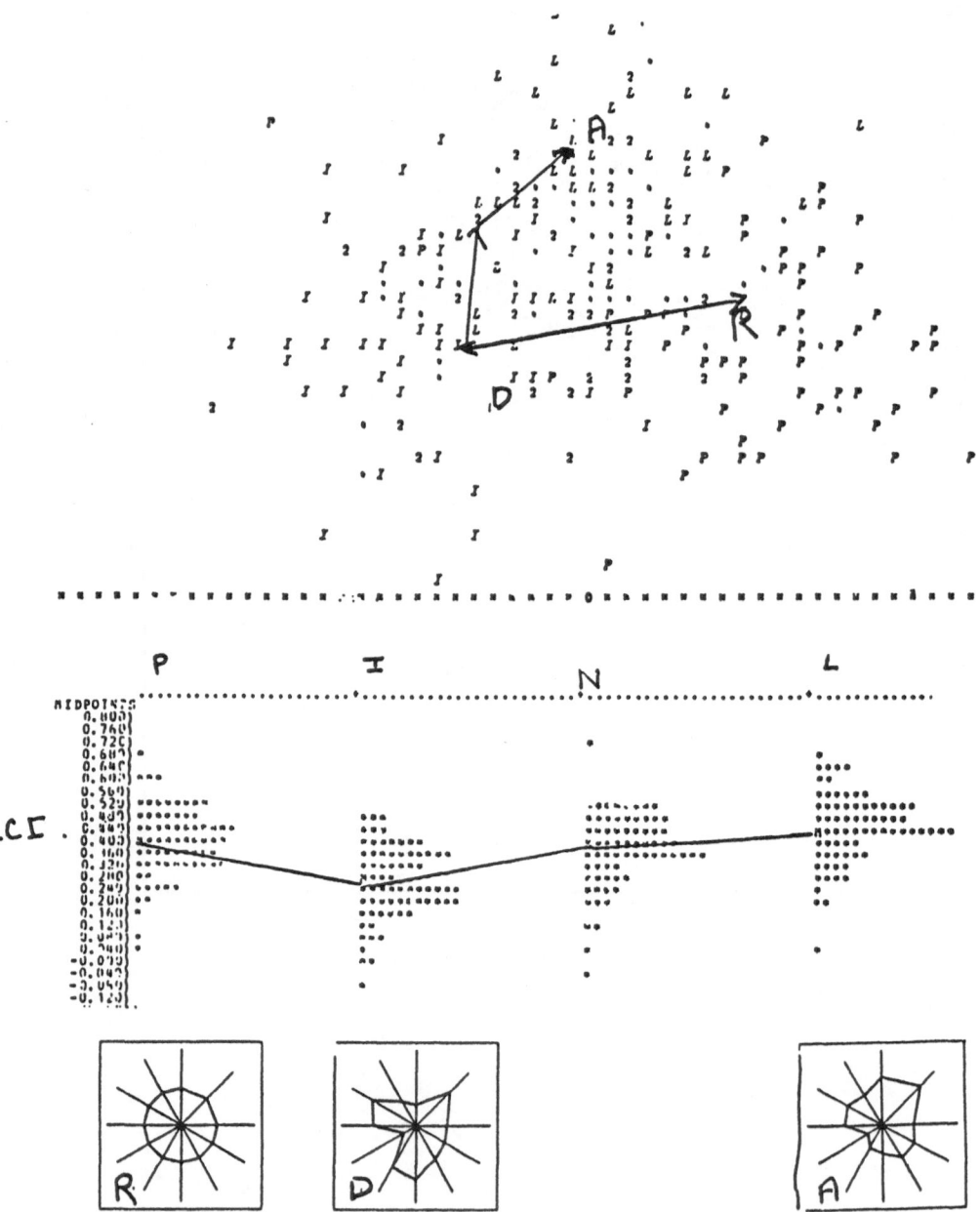

ANALYSIS OF CLINICAL TRANSPLANT DATA: A PERSONAL COMMENT

Gerhard Opelz
Department of Transplantation Immunology, Institute of Immunology
University of Heidelberg
D-6900 Heidelberg, FR Germany

A variety of methods exist for the computation of transplant success rates and their statistical evaluation. For someone not formally trained in statistics (like myself), expressions such as actuarial, regression, multivariate, etc., are impressive and bewildering. Like most transplant physicians, surgeons, or immunologists, I am inclined to respectfully accept p-values delivered by the computer. In my efforts to overcome my statistical deficiencies, one wisdom I adopted without difficulty was that "multivariate" analyses must be superior to "univariate" analyses. After all, the more comparisons and stratifications were carried out the better the result, or so I believed. It was no little surprise, therefore, when M. R. Mickey concluded, after having performed an extensive study based on the UCLA transplant registry data, that multivariate and univariate analyses essentially gave the same answers (1). Had just "some statistician" come to that conclusion, I would not have been bothered too much. That someone who I knew was a renowned member of the international biostatistical community had made the statement, however, did impress me. Could it be that what I had suspected all along but never dared to say was true? That all those complicated computations were necessary only if the data were "no good"? Or to be more precise, that if the patient numbers were sufficiently large, at least with renal transplant data, it did not matter what type of analysis one used? How did that compare with the popular notion that anything other than multivariate was primitive, unsophisticated, perhaps worthless? The revelation that "multivariate" is not unfailingly superior to "univariate" encouraged me to go ahead with writing this manuscript which, I can assure the reader, I was reluctant to do. My apology for contributing the manuscript to a journal of medical informatics is that even naive thoughts sometimes stimulate the experts, if to nothing else perhaps to educate those of us on the "user side".

After listening to many arguments in favor of multivariate analyses, I became fairly convinced that it was time to change my oldfashioned opinion. If indeed new factors could be discovered that hitherto had gone undetected, the improved method obviously should be used. To my own dismay, inspite of considerable efforts employing multivariate models, we ended up confirming what we had known already from our univariate (or oligovariate) studies, rather than learning anything new. In fact, I cannot point to a single example where a factor´s significant influence was discovered first by means of a multivariate analysis. It must be stressed, however, that this statement is valid only for our work with data of the international Collaborative Transplant Study which encompassed some 20.000 transplants when we commenced our comparative studies and which currently has reached over 40.000 transplant records. Just as with simple data subset comparisons, trends of marginal significance obtained with multivariate analyses were sometimes substantiated as the patient numbers grew; more often, however, they proved to be elusive. Over the years, one unsophisticated guideline I learned to follow was that anytime an interesting observation was made in a group of patients numbering fewer than 100 it was not to be trusted. Although I make no claim that this is statistically valid, it pretty much describes the borderline for disappointment in my own work.

I am not unsophisticated enough to discard entirely the usefulness of multivariate analysis methods. Probably, their superiority becomes more apparent if data samples are relatively small. Since our own experience was that they confirmed our previous findings, one could argue that unnecessary work could have been avoided by running multivariate analyses in the first place. Nevertheless, I still maintain that the univariate approach is easier to comprehend and, consequentially, the result is more readily accepted as compared to the rather abstract multivariate computation. Unquestionably, if serious biases or disequilibria existed in the data, the multivariate approach would be far superior. To my knowledge, with the very large series of renal transplants that we have been concerned with, this has not been the case. Even though multiple factors influence the outcome of renal transplants, they tend to be distributed randomly in such a way that univariate analyses usually give correct answers. Perhaps the most important function that I would attribute to multivariate models in our renal transplant setting, therefore, is the provision of assurance that no significant distortions are present in the data. If a finding that was made by single factor comparison is confirmed in a multivariate analysis, one is fairly - although not perfectly - safe in assuming that the answer is correct. Conversely, if the multivariate approach did not support the univariate result, one would have reason for doubt.

An area where we have found the multivariate approach particularly useful is that of outcome prediction. Since the combined influence of multiple factors determines the fate of a transplant, graft outcome can be predicted best by considering all known significant factors according to their individual strength. Following this concept, we developed models that gave excellent correlations of predicted and observed graft success rates (2,3). The result of our most recent effort is shown in Figure 1.

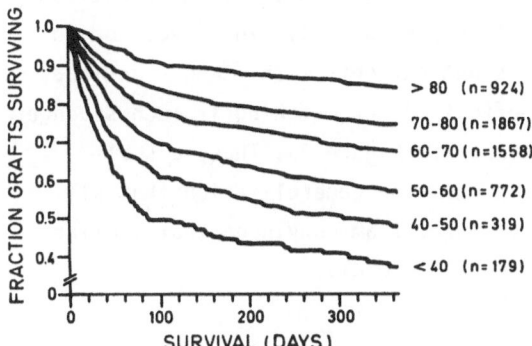

PREDICTED AND OBSERVED SURVIVAL RATES OF CADAVER KIDNEY TRANSPLANTS

With permission of the authors and publishers from Hennige M, Köhler CO, Opelz G: Multivariate prediction model of kidney transplant success rates. Transplantation 42:491-493, 1986.

FIGURE 1. Actuarial cadaver transplant survival curves computed by the method of Kaplan and Meier   for subgroups of recipients for whom likelihood of graft outcome at one year had been predicted in the 9-factor model. Predicted ranges of one-year success rates and numbers of patients studied are indicated at the ends of curves. The prediction model was based on results of 1200 transplants performed in 1982, results shown are those of 5619 transplants done in 1983 and 1984.

Inspite of the prediction model's good correlation with observed transplant results, the model is far from satisfactory for practial reasons. The objective of prediction must be to get a clear choice between likely success and likely failure. Thus, we must strive for a patient distribution that moves as many patients as possible into the extreme categories of excellent or very poor likelihood of success. With the current model, the majority of patients stay in the intermediate categories (Figure 1). Therefore, it is necessary to enhance the model's discriminatory power by adding new variables. Figure 1 was based on nine strong factors: Transplant number, HLA mismatch, blood transfusions, immunosuppressive protocol, lymphocytotoxic antibodies, original disease, general pretransplant evaluation, previous graft duration, and the center effect. Refinements most likely will have to include "weaker" factors and perhaps new factors that have not been recognized in the past.

One of our conclusions was that the factors considered in our multivariate model essentially exerted their effects independently. The same conclusion was reached by M.R. Mickey (1). Nevertheless, I believe there are areas of interdependence that have not been considered sufficiently. It seems likely that improvements in graft outcome prediction can be made by concentrating on the interaction of factors.

The influence of HLA matching in patients with or without preformed lymphocytotoxic antibodies may serve as a simple example. Even though HLA matching and presensitization were found both by Mickey and by us to act independently, the actual data do not support this completely. Someone experienced in transplantation might suspect that patients with high antibody levels might benefit more from matching than patients with low levels or no antibodies at all. Indeed, if one singles out patients with ⟩50 % antibody reactivity against the test panel, the effect of HLA matching is much stronger than in patients with less reactive antibodies (Figure 2). Thus, a small fraction of patients can be identified in whom the generalization that all factors are independent does not apply anymore. I am convinced that similar examples can be found if one examines the data closely.

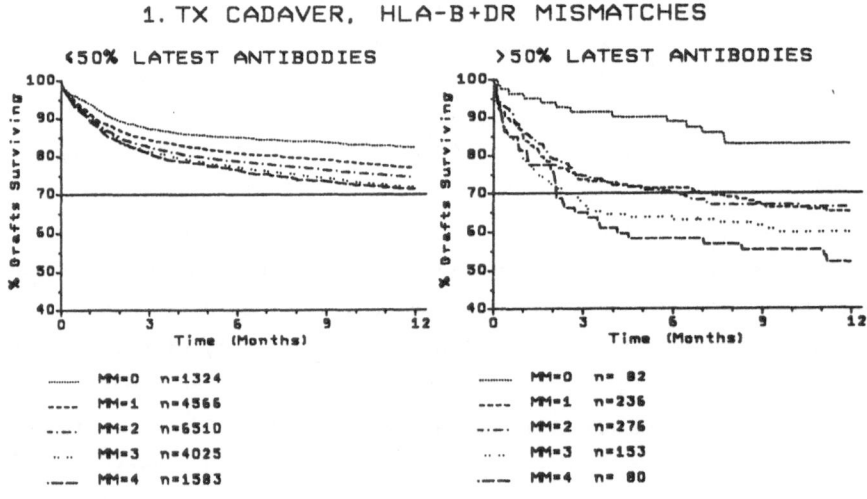

Figure 2. Survival rates of first cadaver kidney transplants. The number of HLA-B and HLA-DR antigens mismatched between donor and recipient was analyzed. The difference between well matched (MM=0) and poorly matched (MM=4) transplants is much larger in patients with ⟩50 % lymphocytotoxic antibody reactivity in their latest pretransplant serum (right half of figure) as compared to patients with ≤50 % antibody reactivity. Numbers of HLA antigens mismatched and numbers of patients studied are indicated for each curve.

An example for true independence is illustrated in Figure 3. Even though the graft success rate is significantly better in patients treated with cyclosporine than in patients not treated with cyclosporine, the effect of HLA matching is essentially the same. In other words, although the success rate is approximately 10 % higher in cyclosporine-treated patients, if one compares individual match grades, the difference in graft survival at one year between transplants with 4 mismatched HLA-B and HLA-DR antigens and transplants with no mismatch is approximately 15 %, regardless of the type of immunosuppression (Figure 3).

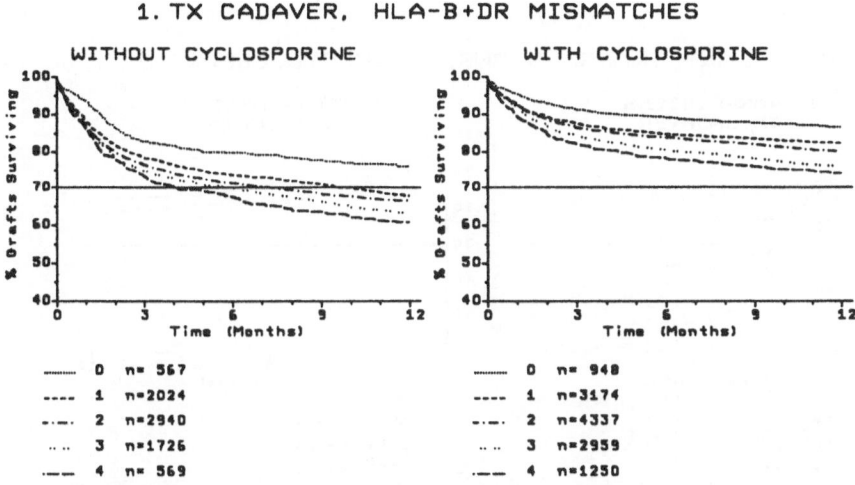

1. TX CADAVER, HLA-B+DR MISMATCHES

Figure 3. Effect of HLA-B and HLA-DR mismatches on outcome of first cadaver kidney transplants. Patients immunosuppressed with or without cyclosporine were analyzed separately. Although the overall survival rate of cyclosporine-treated transplants was 15 % higher, the influence of HLA matching can be seen to be similar in the two categories of immunosuppression. Numbers of mismatched antigens and numbers of patients studied are given for each curve.

A common mistake in the analysis of transplant results is made in the treatment of data that were accumulated over many years. In order to obtain sufficiently large patient samples, data are collected over years and analyzed as one unit. Although this may be appropriate with respect to many questions, there are clear examples where significant changes over time did occur. One such change, which has taken nearly everyone by surprise, is the improvement in the success rate of nontransfused renal transplant recipients. Even though transfused patients had been known for many years to outperform nontransfused ones, there has been a steady improvement of graft outcome in nontransfused recipients. Indeed, the most recent comparison of transplants performed during 1985 shows only a very small advantage of transfused over nontransfused patients (Figure 4). Thus, it would seem appropriate to include year to year variation as an important variable in all future analyses. Combination of material obtained during several years is permissible only if no significant differences in the year to year results are present. In our experience, year to year variation can be ignored for most but not all factors.

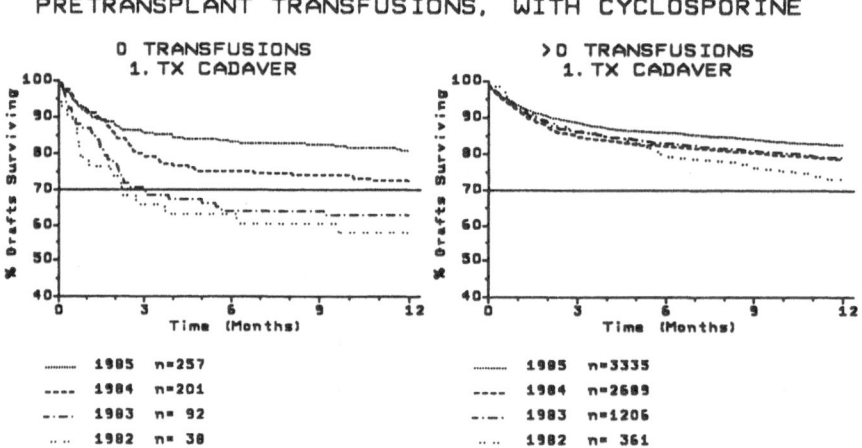

Figure 4. Effect of pretransplant transfusions on outcome of cyclosporine-treated cadaver kidney transplants. Results were computed for transplants performed in 1982, 1983, 1984, or 1985. In patients without pretransplant transfusions (left half of figure) the improvement in graft outcome is much more pronounced than in patients with pretransplant transfusions (right half of figure).

An area which I find difficult to deal with is the relationship between graft and patient survival rates. In renal transplantation, because a patient who rejects a transplant usually returns to hemodialysis, patient survival rates are higher than graft survival rates. A comparison of graft and patient survival for the 3 main categories of donor/recipient histocompatibility is illustrated in Figure 5. As expected, patient survival is better than graft survival in all 3 categories.

RELATIONSHIP

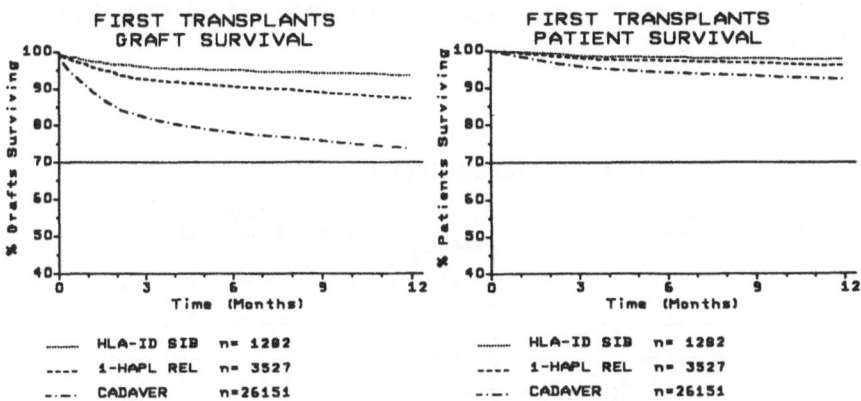

Figure 5. Comparision of graft and patient survival rates in first kidney
transplants. Results were analyzed separately for tissue type identical
transplants (HLA-identical sibling transplants), half-identical related
transplants (HLA 1-haplotype matched transplants, mainly parent to child
transplants), and transplants from cadaver donors. Patient survival rates are
higher than graft survival rates because patients usually return to hemodialysis
if they reject their transplant.

The picture gets more complicated if one examines particular patient subgroups.
For example, if highly presensitized patients are studied, a high rate of graft
rejection is found but patient survival is very good (Figure 6). In contrast, if
one examines patients over the age of 60, one also finds relatively poor graft
survival, however, this is due largely to a high patient death rate (Figure 7).
Thus, although both highly presensitized and elderly recipients have impaired
graft survival rates, the reasons are entirely different. In the first group,
immunologic rejection is the cause, whereas in the second group patient death
accounts for the high failure rate. A common method to adjust for patient death
is to count patients who died with functioning grafts as "lost to follow up".
However, I have found that method unsatisfactory for several reasons. Are we
really sure that immunological factors played no role in "patients dying with
functioning grafts"? And isn't the patient's likelihood to survive more
important than the likelihood to have a functioning graft? Can one really
separate graft from patient survival that easily? Immunologists, are interested
primarily in the influence of immunological factors on graft outcome. Therefore,
studies dealing with the analysis of immunological variables usually are based
on the computation of graft survival rates. Nevertheless, it would be very

helpful if a method could be devised that, without extra effort, gave a "warning signal" every time a low graft survival rate was caused mainly by a decrease in patient survival.

LATEST ANTIBODIES

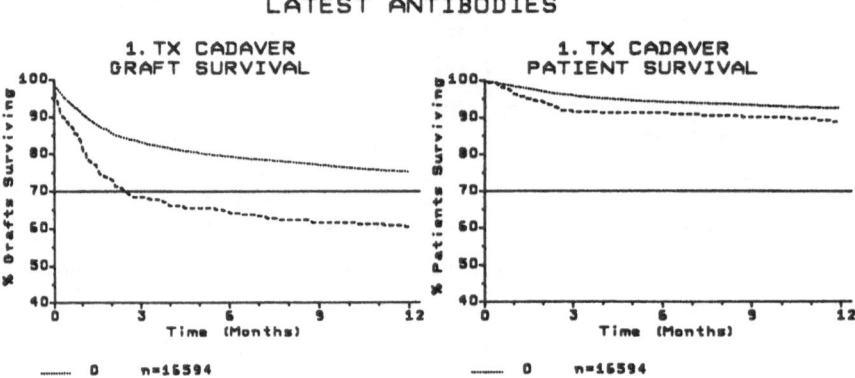

Figure 6. Analysis of graft and patient survival in cadaver transplant recipients depending on their pretransplant antibody status. Patients with highly reactive lymphocytotoxic antibodies show poor graft survival but quite good patient survival as compared to a control group of recipients without pretransplant antibodies.

RECIPIENT AGE

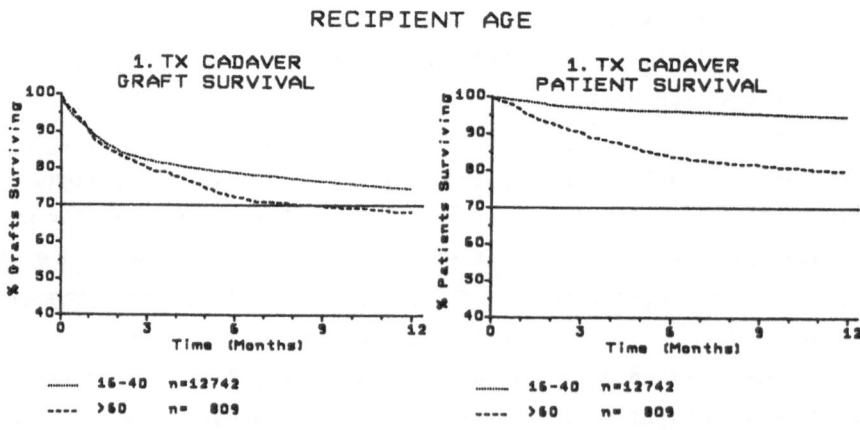

Figure 7. Influence of recipient age on graft and patient survival rates following cadaver kidney transplantation. Patients over the age of 60 have a patient survival rate much inferior to that of a control group of recipients aged 16-40.

Analyses of transplant data are used by clinicians and immunologists to develop improved methods of examining, testing, selecting, and treating their patients. Whether correct or incorrect conclusions are drawn can have dear consequences. Therefore, appropriate methods with the least possible margin of error must be applied. It is also important to devise improved methods of recognizing influential variables as well as interactions among them. With thousands of transplants being performed every year, even small improvements that can be accomplished through sophisticated data analysis must be realized. This personal comment from one on the "user side" is by no means meant as a challenge to established statistical practice. Rather, it is intended as a reminder that our actions must not be directed solely by the noncritical acceptance of p-values.

REFERENCES

1. Mickey MR: Multivariable analysis of one-year graft survival. In Clinical Kidney Transplants 1985 (P.I. Terasaki, Editor), UCLA Tissue Typing Laboratory 1985, pp. 27-44.

2. Mickey MR, Opelz G, Terasaki PI: Prospective estimates of success of kidney transplants. Transplant Proc 11:1914-1915, 1979.

3. Hennige M, Köhler CO, Opelz G: Multivariate prediction model of kidney transplant success rates. Transplantation 42:491-493, 1986.

Design Considerations

for

Transplant Information Systems

R. Engelbrecht

Gesellschaft für Strahlen- und Umweltforschung mbH
Institut für Medizinische Informatik und Systemforschung
Neuherberg, F.R.G.

1. Introduction

Medicine can be characterized as a scientific area with special emphasis on experience, communication and applied research. *Experience* comes from analysis or review of results of a number of diagnosed patients and treatments which are documented in medical records. Every physician has his way of fulfilling this medical duty. Epidemiologists draw conclusions from these medical records; in specific studies they register cases in a controlled way under standardized conditions.

Due to specialization into different medical disciplines there is frequently more than one physician responsible for the patient including the different service units and laboratories. Therefore, *communication* is a fundamental process in medicine, which still relies on telephone, paper, and direct contact for information transfer and authentication. The application of new techniques, therapies, drugs etc. in medical practice is a *research* project continuously developing for decades. Needs come from the demands of quality-of-care and for cost containment.

Beside these demands, or as an expression of them I want to cite three statements given by Opelz at the 1985 Heidelberg meeting on transplantation which may be the "unbroken thread" due to this article.

"The immunologist is expected to provide information on how to best select recipients and donors, when to increase or decrease immunosuppressive treatment depending on the immune systems's reactivity ..."

"Thus, decisions must be based on the relative merits of laboratory findings, which must be viewed against the patient's complex history."

"Sharing knowledge and information among different teams of workers is another important step toward solving a different problem." (G. Opelz, Heidelberg Meeting 1985)

Medical Informatics play a role in all three areas mentioned and could be seen now and in the next future as:

o the support of well understood routine tasks in broad application areas in prevention, diagnostics, therapy, education, and research in health care in an effective and cheap way

o the provision of information, expertise, advice, and knowledge in multiple situations and environments to enhance the quality of care and local problem handling

o the provision of local and regional communication facilities to get patient data, information etc. at various localities, timely and accurate

o the provision of epidemiological and aggregated data for managerial decisions and health care resource planning.

All these tasks are based on methods of medical informatics and information processing and the term "medical information system" may be a synonym for all this in the following chapters. The transplant information system is a subclass of a medical information system with the same basic properties and additional transplantation-specific features.

In general, information systems may be classified according to the types of objects (BLUM 84) they deal with:

o *Data*, uninterpreted items like morbidity distribution in a specific population or patient's blood pressure, name, weight and lab results. They are used for further processing into numerical calculations, reports and decision-making.

o *Information* is context related and gives an interpretation of the data, e.g., a diastolic blood presssure of 90 mmHg is high in the context of WHO.

o *Knowledge* is experience and scientific result of experiments and studies in a formalized way, so that it can be applied to data. It infers information from those data.

The classification is artificial. Because the definitions of all three types are not as sharp as it seems, it is helpful to describe the last three decades of Medical Informatics. It may be said that the 60's focused on data processing, the 70's put more emphasis on database technologies and information processing, and the 80's have started in the field of knowledge processing.

There is a wide range of functions that are performed on these objects to fulfill administrative and medical tasks in primary and hospital care. The problems of the functions are their medical relevance and their scale of integration into the medical routine. Medical Informatics should be measured by its contribution to medical practice and research.

2. Medical Information Systems

A system may be seen as a number of interrelated objects which can belong to different classes of several types. So a system may be described by all the objects or their classes as in the previous chapter. The main object classes are hardware, software, orgware, firmware, teachware etc. and models and functions.

The hardware is that part which is the physical connection to the users. It is the basis for running an information system and may be divided into:

○ _hardware_

○ _computer_, processor and storage, that holds the programs and data and runs it

○ _peripherals_, input/output devices as terminals and printers

○ _networks_, lines and processors to allow communication

○ other _equipment_ as telephones and copy machines.

Developments in this area are most promising for medical computing. The size and prices are reducing so that a wide implementation is to be expected, even in areas such as private practitioner and nursing homes. An example is the dissemination of personal computers since their broad introduction in 1982.

Software tools are programs to construct pieces of information systems on the hardware. They are classified into:

○ Operating systems

○ Programming languages

○ Data base management systems

○ Statistical analysis systems

○ Presentation systems.

The basis is the operating system, which links the hardware to the tools or application programs. Other software components are selectable from a great variety of items; on personal computers there are a number of products, which offer programming languages, data base, analysis and presentation integrated into the same package. So the user has only one environment in which to do his different tasks of information processing. Besides hard- and software, terms as orgware, firmware, teachware, are used. They may be characterized as follows:

○ _orgware_ to assist the information task from an organizational standpoint with analysis, implementation assistance, review

○ _firmware_ is mostly a combination of hard- and software implemented into a chip and contains company specific knowledge, e.g., coding and decoding algorithms

○ _teachware_ consists of courses, implementation guidelines, manuals to enable the user to make optimal use of the information system.

The composition of the different items of hard- and software may be described in three technical models (Fig. 1).

Technical Models of Information Systems

- **Workstation**
  - Intelligent terminal
  - Computer

- **Network of workstations and servers**
  - Local
  - Regional
  - Wide

- **Network of computers**
  - Local
  - Regional
  - Wide

Fig. 1. Technical Models of Information System

In the first model the workstation is the center of interest as a unit containing all the components mentioned above. A typical representative for it is the personal computer.

Due to specialization in the working conditions it is often necessary to specialize the workstation or to link several workstations together to do a meaningful task. Each workstation in the second model is linked to a network which is becoming a common technique in the field of laboratory medicine.

On the other hand a big mainframe computer with a number of more or less specific terminals represents a special category of this second type of assembly.

A network of computers locally or regionally is the third model and is used by large companies, studies, projects or interest groups as for example the European Academic Research Network (EARN). The aim of these different assemblies is to bring adequate information processing power at the right time to the right place. This goal is a key issue of all developments in the field of medical informatics. In some health care areas we are far away from this. A great effort is neccessary from the experts in information technology as well as from many others including statisticians and physicians brought together in expert teams.

A third way to define an information system is by its functions, which are basically the following:

- *information acquisition* is done in a patient-oriented or systematic way, which will be discussed in detail later. This is the most essential task on which all other tasks are constructed

- *information validation* is related to the acquisition process. This is a key to quality and accuracy, and hence for acceptance of the whole system o *information storing* is done in data bases in actual or by archiving files on different media such as tapes, cards, diskettes, and magnetic or optical disks

o *information processing* embraces a wide range of tasks such as calculating body surface from height and weight, classification, coding and decoding, generating reports or continuously monitoring the patient.

o *information evaluation* by statistical methods is the basis of epidemiological and other medical research

o *information retrieval* needs no description. It is the reason for installing information systems

o *information presentation* must be adequate to the problem, for example in a graphical way the user may better perceive the information content

o *information flow control* is necessary to ensure that the information reaches the addressee and privacy is maintained.

Each item discussed above must be considered when designing and implementing a medical information system. Human questions are important, for example in the context of information acquisition it has to be decided whether data entry should be done by the nurse or the physician who are responsible for the data, or by a documentation clerk, who is more familiar with the different entry procedures. An important point is the integration of information systems into the clinical or medical routine. The medical process is described by Reichertz as

"the apperception of signals, the interpretation of data leading to action "(REICH 72).

It is an interactive information process with all the problems of informatics to assist by tools to an enhancement of the quality of care and the reduction of costs. Peterson stated in his opening speech for Medinfo 88, that about 30 percent of all hospital costs are for information processing and studies, and showed that 50 percent of the physicians' workload is documentation and information handling.

The way of assistance in this situation is to bring information processing to the place of work of the user and to deal with new techniques (Fig. 2).

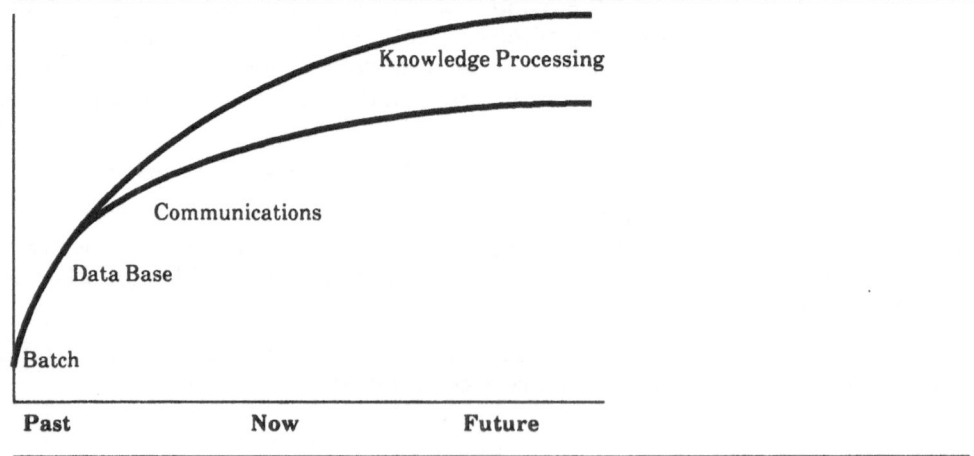

Fig. 2. New Class of Applications (SCHORR 86)

Figure 2 shows a simplified graph of the quality of applications. The development and use of methods of artificial intelligence will add a remarkable element of productivity to the applications.

A key problem in medical informatics is the acquisition of knowledge. Feigenbaum states "The Knowledge Principle simply says if a program is to perform well, it must know a great deal about the 'world' in which it operates" (FEIG 86). This is true for general systems and especially dedicated applications such as transplantation information systems. The process of knowledge derivation can be seen as two stages.

The first includes all routine tasks related to the patient or in other words, the patient serves as an information source (Fig. 3) during his stay in hospital, in an ambulatory care unit or at a private clinic.

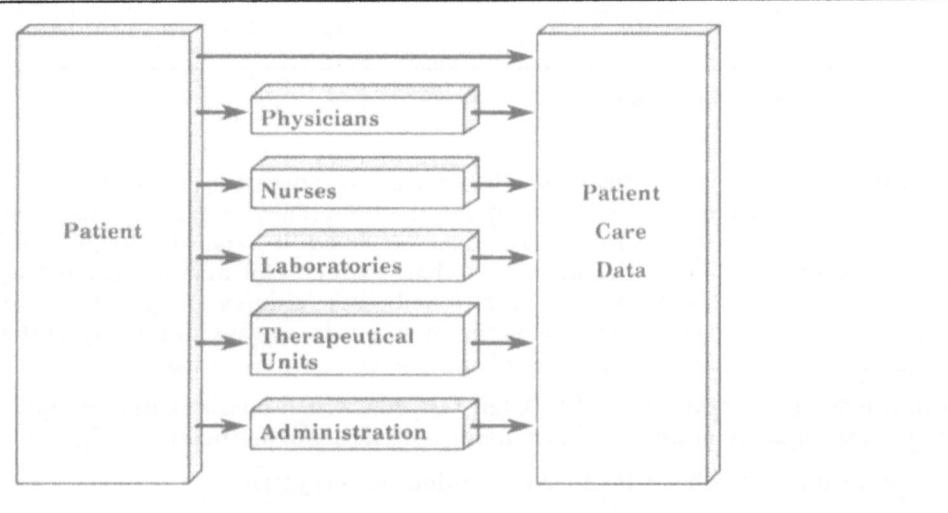

Fig. 3. Facts and Knowledge Derivation

The clinical data are entered directly, interpreted, generated or measured by physicians and nurses: blood pressure, diagnoses, planned therapy, actual medication. Lab results complete this palette. The therapeutic units and administration add further data to this data mosaic stored in medical records and patient-oriented databanks. The complex makes up the medical history.

Evaluation of patient care data, mostly with statistical methods, leads to facts which are stored in a general fact data bank (Fig. 4). Literature and results of experiments are also sources or are stored in total. The general fact data bank is a source for the knowledge bank. Different methods of analysis and transformation are used to create medical knowledge. One process results from experience of the physician. Experiments and literature may be directly entered as knowledge. This is a schematic way of the process of gaining knowledge. The whole research area contributes and it might be better discussed as a network of a large number of small processes containing iterations and sometimes dead ends. Another way of describing the evolution of medical knowledge is proposed by Blum (BLUM 82) in the RX project. Blum shows a cycle of

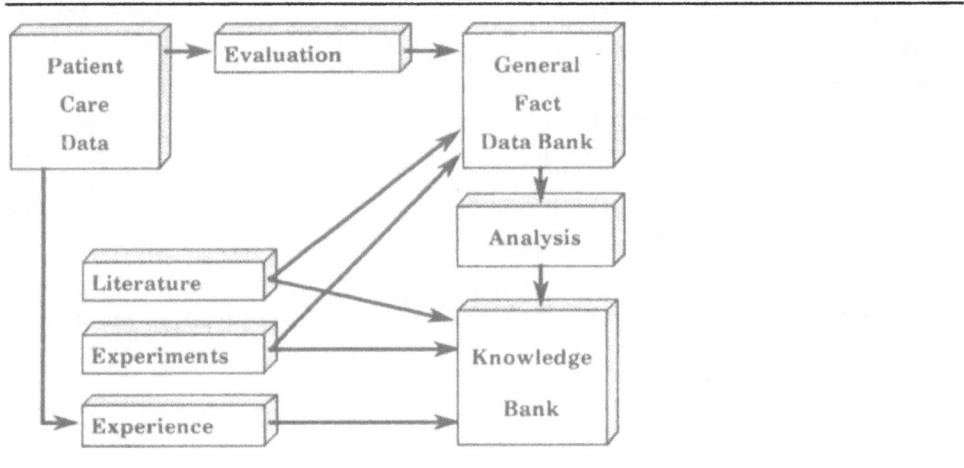

Fig. 4. Facts and Knowledge Data Bank

observations and applying medical knowledge in clinical practice, generates hypotheses to be verified. This leads to new knowledge. In the RX-project this is computer assisted. Cases are taken out of the literature. The discovery module generates hypotheses of a subset of the data base and the knowledge base, which can also be started by the medical researcher. On the basis of these hypotheses studies are generated by the computer system. The entire data base is taken into an evaluation by statistical packages. The results are added as special experience to the knowledge base.

The knowledge generation tool RADIX (BLUM 86), which operates like an epidemiologist, is a consequence of this approach, using techniques of artificial intelligence.

All this work prepares a basis for an overall information system.

The three categories of information systems are:

o *information retrieval using facts*

o *patient information using facts and patient data*

o *medical expert system using facts, patient data and knowledge.*

They may occur in one system and in the future this will be the case. On the other hand an information system may be defined by these three bases: facts, patient data, and knowledge. The terms "system for medical decision making" and "consultation system" have not been used so far. Under some restrictions they may be included in expert systems, because the user, e.g., is only result-oriented and makes no differences. To illustrate the status of development, the facts, the patient data and the knowledge are represented three-dimensionally (Fig. 7).

Hitherto, most medical information systems are to be found in the area spanned by the fact-axis and the patient-data-axis, sometimes only covering the latter. To fulfill all the user requirements it will be necessary to extend the use of facts and incorporate knowledge. An example for this approach is the HELP System. That started as a

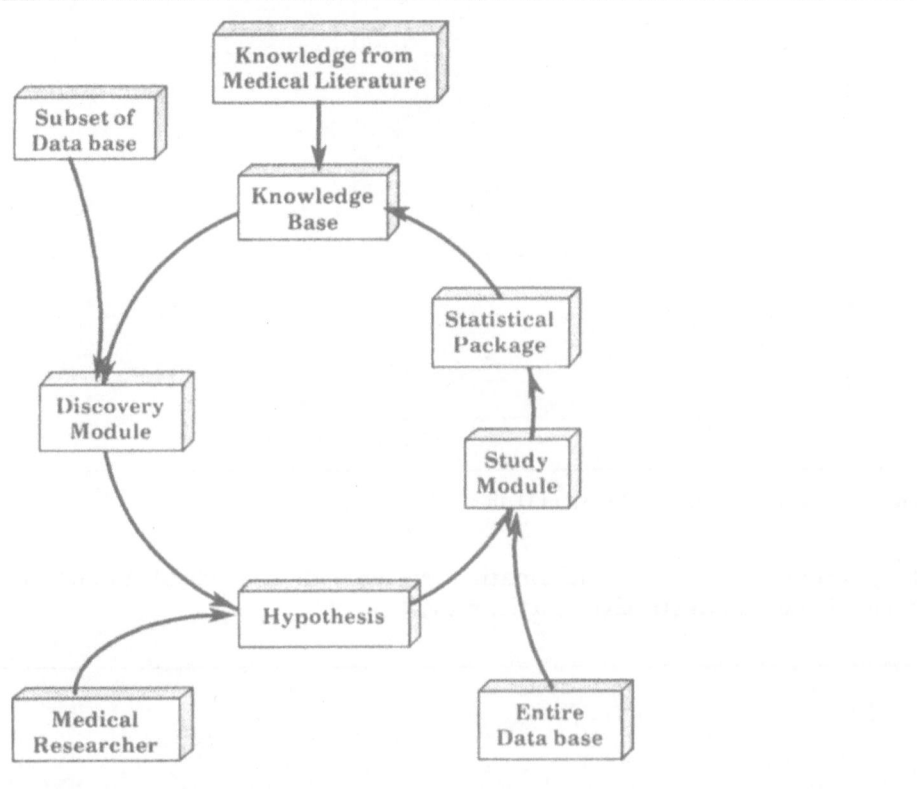

Fig. 5. Discovery and Confirmation in RX (BLUM 82)

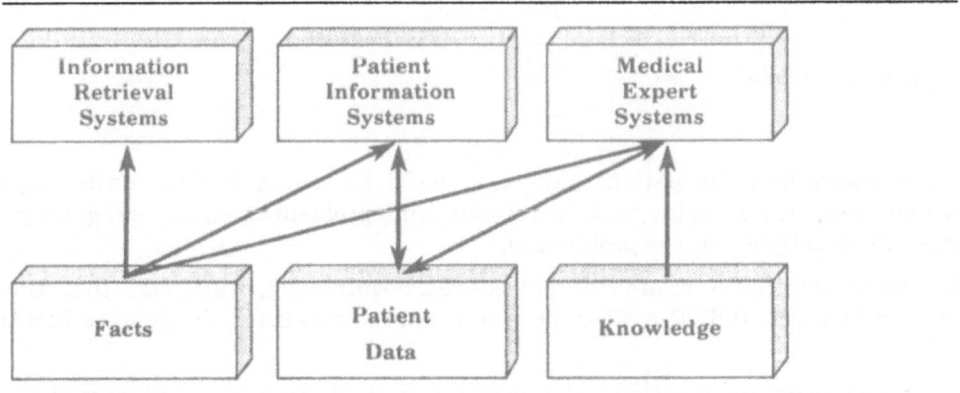

Fig. 6. From Retrieval to Expert Systems

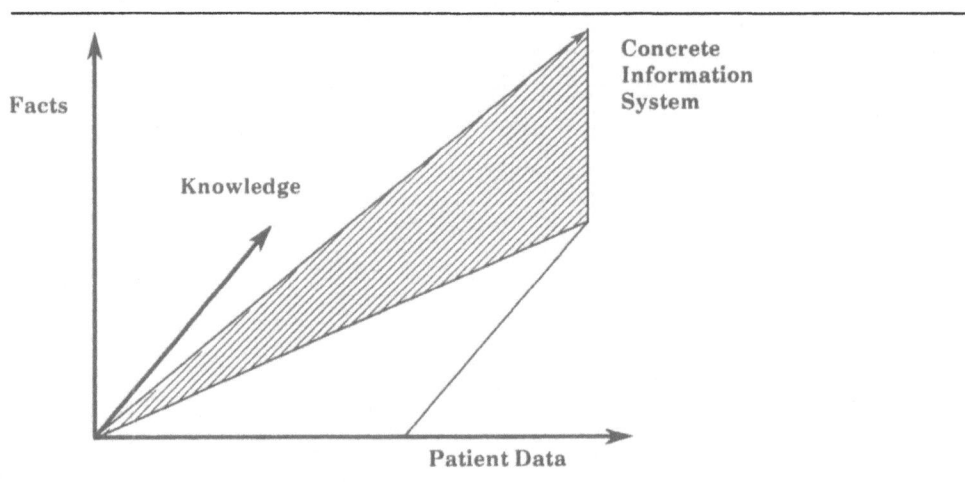

Fig. 7. Facts and Knowledge of Patient Data

medically oriented hospital information system, then medical knowledge was implemented and administrative functions, too.

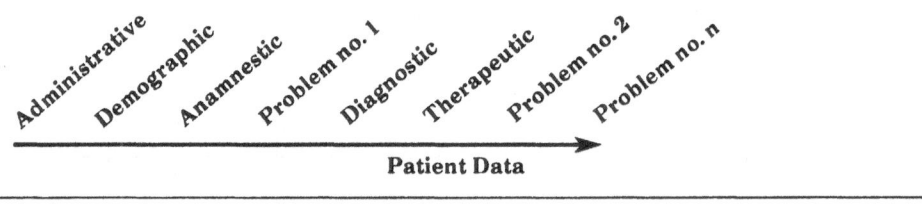

Fig. 8. Patient Data

Figure 8 shows how the patient data axis could be scaled in data units such as administrative, demographic, oral interview and problem-oriented diagnostic and therapeutic data (problem 1 to problem n).

Facts may be classified as standards, procedures, terminology, drugs, etc.. In a study by Santini and Ducrot, 1982 (SANT 82) the information needs for drugs were identified (Fig. 10).

They offered drug information to physicians and pharmacists and recorded the requests. The results show large differences between both user groups in their needs and ranked order of requests. A more detailed subdivision of all the fact data classes is possible and necessary to give a complete picture.

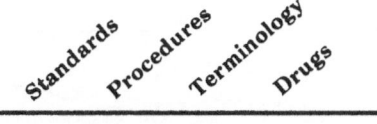

Fig. 9. Facts

Physicians		Pharmacists	
Side effects	30%	Equivalences of foreign drugs	> 30%
Contraindications	15%	Identification of drugs	15%
Interactions	10%	Standard or special use and dosage	10%
		Side effects	10%
		Interactions	6%

Fig. 10. Information Needs for Drug Information (SANT 82)

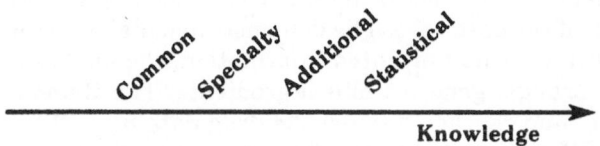

Fig. 11. Knowledge

The units on the knowledge axis (Fig. 11) are "common knowledge", often called "world knowledge", e.g., if someone is pregnant it must be a 'she'. "Specialist knowledge" is that of specialists for internal medicine, transplantation, immunology, etc.. "Additional knowledge" may derive from the field of law or other special, but not medical, areas. "Statistical knowledge" means knowledge of statistical methods and results of statistical evaluations, e.g., epidemiological studies.

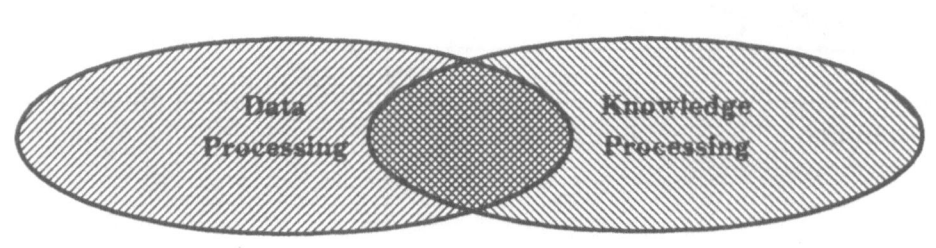

Fig. 12. Integration (SCHORR 86)

Integration (Fig. 12) of conventional patient information systems and expert systems for data and knowledge processing will occur, as was stressed by Herbert Schorr at the 86 AAAI (the annual meeting of the American Association of Artificial Intelligence). That will be a combined challenge for medical informatics and medicine itself.

An example for an information system which deals with facts and expert knowledge is the SMA (Scholz-Medis-Arzneimittelinformationssystem), a drug information system, should be explained in a typical application. The system gives information on drugs and their composition, on ingredients and ingredient groups including all drugs and synonymes contained as well as on interactions between the drugs contained in a prescription. The system indicates interactions which are depending on the route and use including the corresponding appropriate dosage. The effects and the mechanism are explained, alternative drugs are listed and recommendations are given. It is possible to substitute the drugs of the prescription for others and to check again for interactions. The access to information about indication, contraindication, side effects and dosage of drugs has already been realized conceptionally. At the moment this part of the data base is in the process of being established.

An extension of the data base by further information regarding pharmacodynamics, pharmacokinetics and technical details like price, different sizes of the packets is being prepared as well as the integration of the data base. The demonstrated dialogue consists of four typical screens. In the first the patient oriented information, the medication or prescription, is entered as set of brands, generics and ingredients. The items of the prescription are entered separately and checked against the drug data base. When the full prescription is complete, the SMA is started to test for interaction. Interactions in the SMA are defined between 2, 3 or 4 drugs, drug groups or ingredients, that means the prescription is decomposed into ingredients and classified in groups.

The result of the prescription of Fig. 13 is shown in Fig. 14. For each interaction the interacting items are shown and the interaction is classified into effect, frequency, and significance.

```
ID.-Feld        SCHOLZ - MEDIS - Arzneimittelinformationssystem      27.08.86
WEIN22          *****  S M A - Version 2.2 Lizenz APOTHEKE  *****        12:12
------------------------------------------------------------------------------
Wechselwirkungen                    ** REZEPTUR **

Nr Praeparat/Wirkstoff              Form Applikation
-- ----------------------           ----  ------------------
 1 Marcumar                         Tbl  syst. enteral
 2 Euglucon N                       Tbl  syst. enteral
 3 Dolviran                         Tbl  syst. enteral
 4
 5
 6
 7
 8
 9
10

Praeparat/Wirkstoff:
                                                            Funktion ==>  __
--------------------------- F u n k t i o n e n ------------------------------
AR Alte Rezeptur          TW Test auf Wechselwirkung       HM Hauptmenue
NR Neue Rezeptur          SF Suche Fertigarzneimittel
ST Storno                 SW Suche Wirkstoff
```

Fig. 13. Prescription to be tested for interactions and the ingredient

```
ID.-Feld        SCHOLZ - MEDIS - Arzneimittelinformationssystem      27.08.86
WWLS21          *****  S M A - Version 2.2 Lizenz APOTHEKE  *****        12:12
------------------------------------------------------------------------------
Wechselwirkungen       ** UEBERSICHT: 3 Wechselwirkung(en) gefunden **

Nr Praeparate          Ausloesende Wirkstoffe       Ap   ED (mg)     TD (mg)
-- ------------------   ----------------------       --   ---------   ---------
 1 Marcumar             Phenprocoumon                S*
   Euglucon N           Glibenclamid                 S*
   .......... / kein Eff. / wahrsch. / -                          / ..........

 2 Marcumar             Phenprocoumon                S*
   Dolviran             Acetylsalicylsaeure          S* >  250.000 >   40.000
   .......... / gefaehrl. / selten / Blutungsneigung erhoeht   / ..........

 3 Euglucon N           Glibenclamid                 S*
   Dolviran             Acetylsalicylsaeure          S*          > 3000.000
   .......... / bedeuts. / wahrsch. / Hypoglykaemie            / ..........

Auswahl Nr:  __                                        Funktion ==> EF
--------------------------- F u n k t i o n e n ------------------------------
SU Substitution        EF Effekt.Risikofaktoren      HM Hauptmenue
                       ME Mechanismus                RU Ruecksprung
                       VO Vorschlag                  +- Blaettern
```

Fig. 14. Interactions found for the prescription in Fig. 13 with the interaction partners in each first two lines and short description in the third line

```
ID.-Feld        SCHOLZ - MEDIS - Arzneimittelinformationssystem        27.08.86
WWME21        ***** S M A - Version 2.2 Lizenz APOTHEKE *****           12:12
--------------------------------------------------------------------------------
Wechselwirkungen              ** MECHANISMUS **

Praeparate              Ausloesende Wirkstoffe           Ap  ED (mg)    TD (mg)
----------------------  ------------------------------   --  ---------  ---------
Marcumar                Phenprocoumon                    S*
Dolviran                Acetylsalicylsaeure              S* >  250.000 >   40.000



Drei Qualitaeten  der Acetylsalicylsaeure nehmen  auf das  Blutungsrisiko Ein-
fluss: Thrombocytenaggregationshemmung,  Magenschleimhautschaedigung, Einfluss
auf die Prothrombinaktivitaet.
Acetylsalicylsaeure hemmt (irreversibel durch  Acetylierung der Cyclooxygenase
und daher im Gegensatz zu nicht  O-acetylierten Salicylaten) schon bei niedri-
ger Einmaldosis  von 250mg oder chronisch  40mg/die die Synthese  des Prostag-
landins Thromboxan A2, das die Thrombocytenaggregation ausloest. In Abhaengig-
                                                          Funktion ==> VO
--------------------------- F u n k t i o n e n ---------------------------
SU Substitution          EF Effekt, Risikofaktoren             HM Hauptmenue
                                                               RU Ruecksprung
                         VO Vorschlag                          +- Blaettern
```

Fig. 15. Description of the mechanism of interaction No. 2 (Fig. 13)

```
ID.-Feld        SCHOLZ - MEDIS - Arzneimittelinformationssystem        27.08.86
WWSU21        ***** S M A - Version 2.2 Lizenz APOTHEKE *****           12:12
--------------------------------------------------------------------------------
Wechselwirkungen              ** SUBSTITUTION **

  10 Marcumar                              20 Dolviran
     Enthaltene Wirkstoffe:                   Enthaltene Wirkstoffe:
 >11 Phenprocoumon                         >21 Acetylsalicylsaeure
  12                                        22 Codein
  13                                        23 Coffein
  14                                        24
  15                                        25

------------ Moegliche Substitute fuer Praeparate bzw. Wirkstoffe ------------
  31 Paracetamol                            34 Carprofen
  32 Sulindac                               35
  33 Ibuprofen                              36

Substituend (Nr.)  : __
Substitut (Nr./Name): __                              Funktion ==> TW
--------------------------- F u n k t i o n e n ---------------------------
TW Test auf Wechselwirkung EF Effekt, Risikofaktoren        HM Hauptmenue
   (mit Substitut in R.)   ME Mechanismus                   RU Ruecksprung
ES Eintrag Substitut in R. VO Vorschlag
```

Fig. 16. Recommended substitutes to avoid the interaction or reduce it for the interaction partner Dolviran and the ingredient Acetylsalicylsäure

The consultation component after the detection of one or more interaction is predefined within SMA. Two typical questions in the case of interaction are, how does this interaction work and what can be done to avoid this interaction or to minimize it, if their is no way without these drug combination.

The specific mechanism of the interaction between Marcumar and Dolviran is shown in Fig. 15 and contains the German text in this version. A recommendation in full text is available, which in one ore more screens gives in this specific case regulations and alternative drugs.

The alternative drugs are listed on the screen called 'Substitution' (Fig. 16) and it is possible to select one ingredient as a substitute, e.g., Paracetamol and the system retrieves for drugs with Paracetamol and without Acetylsalicylsäure (Acetylsalicyl Acid). The user has to select, and the system will test for interactions again. At least the integration of patient data or the integration into a patient information system with indication, contraindication, and risk factors, will provide for the full power of drug consultation.

A special component for patients with renal failure, patients under hemodialysis, and patients after transplantation is under development.

3. Transplantation as a process

The other way of classifying information systems according to their content under the three categories is mentioned in the previous chapter. SMA in this sence is a pharmaceutical system. Medical patient-oriented systems when being directed towards a specialty then they are called departmental systems, or when capturing data of patients with specific diseases or events then they belong to the category registries. Typical examples are cancer registers. A transplantation information system fulfills both descriptions. In some areas it may monitor and assist within the transplantation process; on the other hand it captures only data from specific patients: the recipients and the donors.

Registries are starting with an event. E.g., the MONICA-project, - a ten years' study of "Monitoring of trends and determinants of cardiovascular diseases" coordinated by the WHO and carried out by 42 centres all over the world, - has as an event a myocardial infarction, which may occur several times per patient. There is a clear definition what an infarction is and additional data are gathered at predefined periods of time. Typical for most medical information systems is the time orientation. This might become clear when the transplantation process is seen in these single phases as it is shown roughly in Fig. 17. Starting point is the detection of a malfunction of the organ , e.g., the kidney and a first basic anamnesis. After some time a dialysis may be necessary. This phase (Fig. 17) is accompanied by more and different laboratory tests every 3 months additionally to the routinely done tests. Phase 2 may be called the pretransplantation phase and lasts on the average 2 or 3 years with several tests and examinations at least every three months.

Renal Disease and Transplantation Process and Phases

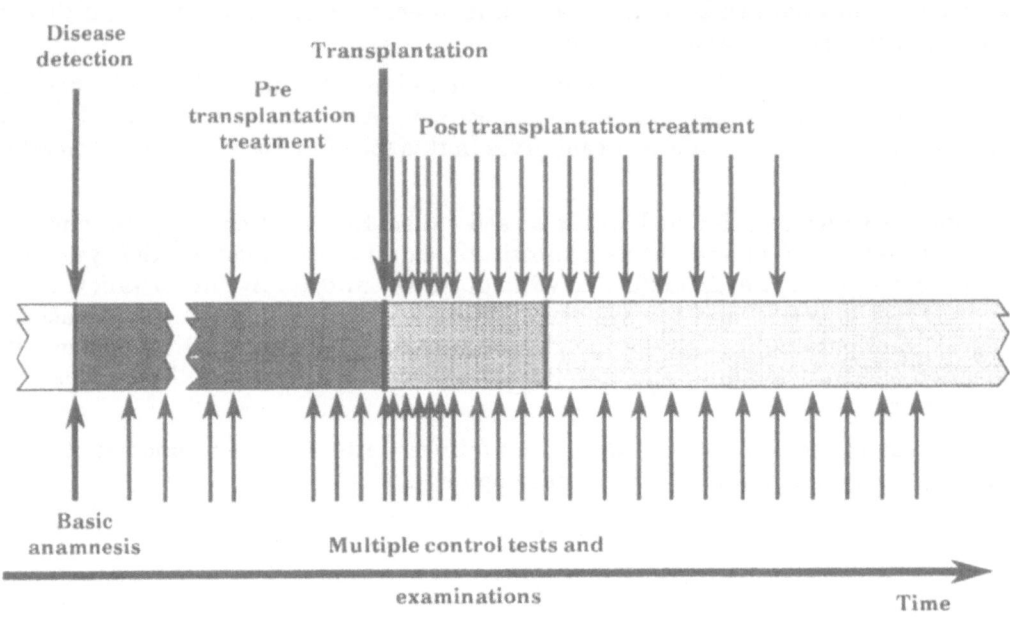

Fig. 17. Renal Disease and Transplantation Process and Phases

Phase 1 **Dialysis** **> 2 years**		
Tests every 3 month		
Phase 2 **Pre transplantation treatment**		
Tests < every 3 month		
Phase 3 **Clinical transplantation**		**2 weeks**
Phase 4 **Post transplantation treatment**		
1st year: 3 exams/week to 1 exam/month		
2nd year: exams every 3 month		

The most intensive phase in treatment and data measurement and gathering is the clinical transplantation phase, whose duration under normal circumstances is about 2 weeks. After that, in phase 4, the intensity of data is going down to one set of examination data every three months. The phases 2 to 4 may occur more than once in case another transplantation becomes necessary. From the standpoint of data processing there are two problems in this type of time-oriented data. The first one deals with the problem of data structure, which might be different in every phase. But at least there will be a minimal basic set which is consistent over all the years. Longitudinal evaluations can only be done on these data. But an evaluation over time rises the second problem: not all time distances are the same and so the data are not usable for several biostatistical procedures without preprocessing. Data evaluation is necessary for generating medical knowledge, as demonstrated previously. For organizational purposes the data structure looks different for each phase and at least each user.

The data base is the backbone of an information system. It is from the user's point of view not necessary which data model is represented, e.g., relational, hierarchical or network, if he has his model of the data base and is able to work with it. Therefore, a general discussion on data models may be helpful, when the user has specified his needs and his data. The above mentioned MONICA-project may serve as an example for the relation of data, their structure and an organizational solution in hard- and software.

The aim of this project is to gather information about incidence, letality and mortality of myocardial infarction and to get a risk profile. Beside this, information about the health services and their changes is of importance in order to evaluate their influence on mortality. The project is planned for ten years containing 3 independent surveys, a cohort's study and a register study with a yearly follow-up.

The register is designed to store all incidences and letal cases of infarction of the study area of Augsburg, a city 50 km away from Munich and will include on the average 1000 cases, which are between 25 - 74 years old. The data are of equal importance for the organization of the study and their evaluation. Figure 18 shows a rough classification of the data with some examples in table 1. Personal data together with parts of the medical data and the study management data are forming the basis for a complete and accurate gathering of the single cases. The medical and demographic data and the generated core data are the basis of the evaluation of the study and necessary to reach the objectives: generation of high quality and international comparable information for the monitoring of cardiovascular diseases, which is collected on a WHO-basis in the data center at Helsinki.

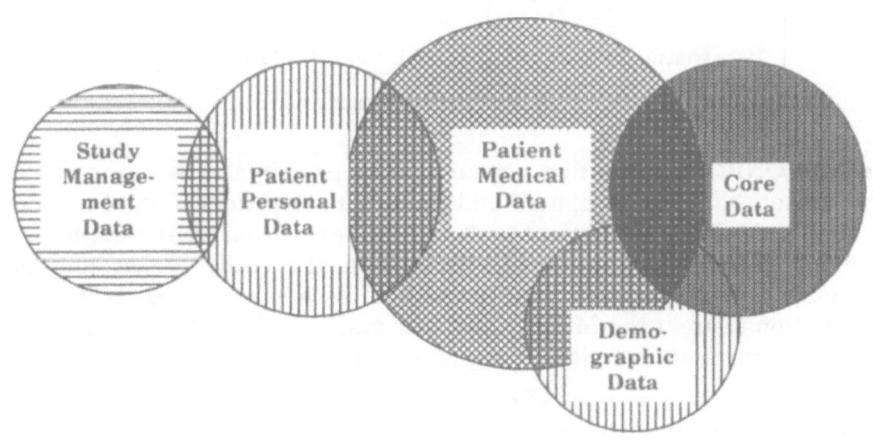

Fig. 18. MONICA-Data Concept (structure)

Table 1. MONICA-Data Concept (elements)

Study Management Data	Patient Personal Data	Patient Medical Data	Demographic Data	Core Data
Hospitals	Name	Diagnosis	Age distribution	CV-classifi-
Physicians	Adress	Anamnesis	Sociological	cation
...	Birthday	Drugs used	structure	ECG
...	...	Lab results	...	...
	...	...	...	...
		...		

Table 2. MONICA-Information Processing Concept

Data Area	Location	Hardware	Software
Patient personal	Study Center Augsburg	IBM-PC	PC-DOS
Study management			Knowledge Man (DBMS) WORD (Textprocessing)
Patient medical	MEDIS-Institute	IBM-4381	VM/CMS
Demographic			ADABAS/NATURAL SAS
Core	Data Center Helsinki	VAX 11/780	VMS

The concept for the MONICA-information system has two principles: the storage of the data in that place where they are computed and the optimal consideration of privacy. Table 2 reflects these principles. All patient personal data are stored on an autonomous computer system, a personal computer (PC) located in the study center and not accessible from outside. The complete scientific medical data are case-oriented transported to the evaluation team at the MEDIS-Institute and stored in an ADABAS-data base. They are usable for evaluation without any restriction to privacy reasons.

The specific possibilities of PCs were used in designing the local information system. For the ease of use the handling of the system is mouse-oriented and, e.g., the selection and printing of working lists is done by simple "mousing". The lists are usable for immediate further data gathering and checks. The result is an optimal data quality caused through numerous checks for validity and completeness.

In total, the local MONICA PC-based information system supports functions described in figure 19.

Functions of the local MONICA information system

- **Register**
  - Case input
  - Update
  - Interactive queries

- **Organization**
  - Working lists for
    - Overview
    - Control
    - Medical record order
    - ECG-coding
  - Reports for
    - Participating physicians (private and in hospitals)
    - Health authorities
    - Evaluation

- **Text processing**
  - Letters
  - Documentation
  - Publications

Fig. 19. Functions of the local MONICA information system

These functions are comparable to those which must be fulfilled in transplant information systems.

Another short example may be discussed, which is more equivalent with regard to medical specialty, because it deals with transplantation of the bone marrow.

The University of Leiden offers an information system, "The Leiden Data base for BMT Patients", which is distributed to all interested hospitals all over Europe. This guarantees a uniform handling of data and in that way an optimum in comparability. Responsible for this system core are the departments of hematology and pediatrics. The structure of this data base is shown in Fig. 20 and gives a rough idea about the distribution of the 705 fields, which are stored per patient as a maximum. 138 of the data fields contain dates. That shows that this data base is time-oriented in several segments. Problems of this were discussed earlier. Most of the fields, in total 456 per patient, are parameters describing the patient's status in the form of numeric values. For this part of the data base standards are used as the United Nations' country code, the classification of diseases and the classification of micro-organisms compatible with the "International Bone Marrow Transplant Registry". The scope of the data is comparable to the "International Information System on Kidney Transplantation" (Fig. 21) as described by Hennige (HENN 82), where more details can be found.

The Leiden Database for BMT Patients

Responsible: Depts for Paediatrics and Hematology,
Univ. of Leiden

Structure: **General information**
- Reports
- Patient (recipient)
- Donor

Status of primary disease
- At diagnosis
- Prior to conditioning

Treatment prior to transplant
- Antimicrobial
- Immunosuppressive
- Supportive

Transplantation
- Engraftment
- Response of disease

Treatment after BMT
- Supportion
- Antitumor

Complications

Fig. 20. The Leiden Database for BMT Patient

International Information System on Kidney Transplantation*)

Data groups per transplantation:
- Personal data of the recipient
- Medical data of the recipient
- Personal data of the donor
- Medical data of the donor
- Data concerning graft and transplantation
- Data about immunosuppressive treatment
- Laboratory data
- Follow-up data
- Diagnosis of posttransplantant malignancies

*) Hennige 1985

Fig. 21. International Information System on Kidney Transplantation

4. Design Objectives

The key issues for the development of medical information systems were discussed in the last chapters and result in medical relevance of the gathered data, and their use for evaluating the health system, monitoring medical processes, and assisting in the cases of transplantation. But these objectives are only some of the points (Fig. 22), which are relevant to consider when designing and building information systems in the field of medicine.

An information system is to be seen as a tool to improve medicine as an automated assistant. But the only way to reach this aim is to integrate the system into the medical process so that extra effort is not needed to enter data into the information system and to retrieve information. It must be useful in all its components and all components must be integrated so that, e.g. data and their description from the data base are easily transferred to the evaluation part of the system without unnecessary manual action, e.g., for a statistical package.

This is of great importance and must be defined in the early stage of system design, because a system is never built in one step and by only one person. This includes a stepwise implementation into the medical practice and must be planned as a smooth emigration from the conventional system to the automated version. The medical information system is an image of the related medical world. Therefore, it has to be assured that the system is as consistant and as correct as its model. This is one condition

---

Design Objectives of Medical Information Systems

- **Integration**
    - ◦ Into the medical process
    - ◦ Of all system components

- **Smooth migration**

- **Assurance of consistancy and correctness of the (medical) system**

- **Dynamic expansion**

- **Ease of use**

- **Understandibility by the user**

- **Information center**

- **Decision support**

- **Process control**

Fig. 22. Design objectives of medical information systems

for a success and a step on the way of understanding by the user and is one basis for the ease of use of a system. Other criteria for the user interface are discussed separately.

The medical system is a living system and in that sense an information system must be adaptable to the changes of the real world in an easy way and expandable to new functions and user requirements dynamically. The information center concept provides the user with tools to develop functions by his own and to complete and adapt the medical information system.

Special emphasis for the acceptance of a system is to be given to the user interface. It must fit into the daily situations a user has to deal with, to help and to assist him in an adequate manner whenever he is in trouble or uncertain about the possibilities of the system. There are lots of experiences in this field (ENGEL 86) and some of the construction rules are listed in Figure 23. Most of them are self-explicable and they should not be discussed here.

Construction Rules for the User Interface

- **Unique through the whole system**

- **Short response times**

- **Easy to handle**

- **Easy to learn**

- **Usable without special DP-knowledge**

- **At every time the user should know**
  - what the status of the system is
  - what function he can perform
  - what the system expects him to do

- **At every time the system should know**
  - what reaction the user may have
  - what problems the user may have

- **The system should have a model of the user**

- **The system should be intelligent**

- **The system should react similar in similar situations**

Fig. 23. Construction Rules for the User Interface

Conclusion

Medicine is getting more and more complex and even the specialist physician will get more and more spezialized. The medical care is based on experience, experience which is made by the physician himself or others during treatment or research activities. The medical informatic is able to assist the medicine in these processes with the design development, and provision of information, and communication systems. During the past years advantages and disadvantages have shown how and where to implement medical information systems with success.

The new technologies in data processing offer a great chance for complex systems with knowledge-based components for decision support and consultation. These systems are based on medical research as well as on informatic research. Both scientific disciplines are partners in this case and will influence each other in a fruitful manner.

Building information systems in the field of kidney transplantation is an interactive process between physician and information. In a first step the requirements are defined and a system for gathering data is implemented. This is the basis for evaluation and epidemiological research. The results of this research are able to rise more understanding for medical processes and lead to better medical services.

References

BLUM 84   Blum, B.I.: Why AI. The Eight Annual Symposium on Computer Applications in Medical Care, Cohen, G.S. (ed.). Washington, D.C.: IEEE Comp. Soc. Press 1986, 3-9.

BLUM 82   Blum, R.L.: Discovery and Representation of Causal Relationships from a Large Time-Oriented Clinical Database: The RX Project. Lecture Notes in Medical Informatics (19), Reichertz, P.L., Lindberg, D.A.B. (eds.). Berlin - Heidelberg - New York: Springer 1982.

BLUM 86   Blum, R.L., Walker, M.G.: Towards Automated Discovery from Clinical Databases: The RADIX Project. MEDINFO 86, Salamon, R., Blum, B. Jorgensen, M. (eds.). Amsterdam - New York - Oxford: North-Holland 1986, 32-36.

ENGEL 86   Engelbrecht, R.: Experiences in Designing Human Computer Interfaces for Doctors' Office Computer. Human Computer Communications in Health Care, Peterson, H.E., Schneider, W. (eds.). Amsterdam - New York - Oxford: North-Holland 1986, 219-239.

FEIG 86   Feigenbaum, E.A.: Autoknowledge: From File Servers to Knowledge Servers. MEDINFO 86, Salomon R. Blum, B. Jorgensen (eds.). Amsterdam - New York - Oxford: North-Holland 1986, xLiii-xLvi.

HENN 85   Hennige, M.: Implementation of an International Information System on Kidney Transplantation. Meth. Inform. Med. (24) 1985, 135-140.

REICH 72   Reichertz, P.L.: Summary Address: Analysis and Concept. IBM Medical Symposium, Heidelberg, September 1972.

SANT 82   Santini, C., Ducrot, H.: Problems arising from the experience of a drug data bank: collection of data, ways of interrogation. The impact of computer technology on drug information, Manell, P., Johansson, S.G. (eds.). Amsterdam - New York - Oxford:North-Holland 1982, 57-62.

SCHORR 86   Schorr, H.: AI: The Second Wave. Fifth National Conference on Artificial Intelligence, August 11-15, 1986, Philadelpia.

STANDARDISATION OF DATA IN TRANSPLANT REGISTRIES

N.H. Selwood, B.Sc., Ph.D.
Deputy Director,
U.K. Transplant Service,
Southmead Road, Bristol BS10 5ND.

Technical Director,
E.D.T.A. Registry,
St. Thomas' Hospital, LONDON SE1 7EH.

Introduction

The need and scope for standardisation within a data registry often appears to
be self-evident and is usually expressed in terms of efficiency, reliability
and in easing the problems of collaboration.

There are some identifiable disadvantages also to standardisation depending on
the nature and extent of implementation. The balance of the equation of
benefits and drawbacks needs to be carefully evaluated in a highly practical
way lest the advantages turn out to be smaller in comparison to the
disadvantages.

The Practical Situation

The process of standardisation of data within any transplant registry
commences at its inception with the decision of what the main function or
functions of the registry will be, and which aspects of those functions are to
be supported by data gathering or data generating activities. Not all
transplant registries have the same purpose and, whilst there are inevitably
some areas of overlap there are clear differences. Figure 1 shows the example
of the relationship between Collaborative Transplant Study (CTS), the European
Dialysis and Transplant Association (EDTA) Registry and the organ exchange
organisations. These organisations have research interests in common and have
extensive overlap in the data they gather but are nonetheless, functionally
different.

Clearly, the data gathering activities should be limited to harvesting only
data necessary to support the functional activities. However, transplant
registries tend to move the boundaries of their interest and consequently the
data requirements. The trend for evolution will be discussed shortly since it
has important implications for standardisation practices. In the context of a

discussion on standardisation, it should be borne in mind that a single transplanting centre can legitimately be regarded as a transplant registry. It is self-evident that, where a registry generates all of the data itself and functions in complete isolation (which is a fairly unlikely event whilst case numbers are small), problems associated with standardisation are minimal. In the case of a single registry, the simple expedient of a manual document or form for data gathering establishes some standardisation; format, data content, definition and structure can all be established in this way. Even in this simplest of cases however, standardisation may be imperfect since incomplete data can occur and might even be unavoidable.

There are, of course, a variety of reasons for standardisation of data and several ways of achieving it. There are also problems and difficulties, in this discussion these are considered in a practical context.

Why Standardise

The greater the extent of standardisation the greater the expectations of a registry's data handling activities. Most registries would agree that well-defined, accurate and complete data simplifies matters considerably. Also such ideal material can only enhance the statistical validity of analyses performed on it.

The objective of statistical validity is without doubt the best reason for standardisation. A data set which has been rigorously defined to remove ambiguity and in which missing data has been minimised or abolished altogether, together with a structured protocol to provide the basis for definitive analysis of any type and of any orientation, ought to be the goal of every registry.

A process of standardisation which is progressively refined can be a useful tool to first model and later to determine a general framework for data structures which are effective and efficient.

A very commonly perceived role for standardisation is in the sphere of collaboration with other registries or workers. In this context, comparability of data achieved through standardisation is of great importance. Without comparability, the value of collaborative analyses diminishes very rapidly. Care is needed however when implementing standardisation in this context since, as will be discussed below, it can impose practical restrictions. Where registries are computerised it may be advantageous to consider a less rigid approach as will be discussed later. The matter of rationalisation of data gathering activities is assuredly a very important one which is often not well considered and which, with modern data handling capabilities, could and should be greatly improved over the present situation.

Consideration in the light of these needs, and the relative roles that they have in individual registries, can be given to what data might be subjected to standardisation.

What to Standardise

The question of what to standardise can be considered from two aspects, the data content and data structure. Figure 2 shows some typical headings, relevant in transplant registries, that could be considered for standardisation. Each would need to be considered separately in the relevant practical and functional contexts.

Clearly, in a research context and for a truly rigorous study, optimal standardisation of the total data set would be invaluable. However, given that functional fluidity exists in most registries, this may be unachievable except in the context of special studies, and less ambitious but practical objectives should be considered.

At all times the use to which data is to be put in the functional context of the registry, must be borne in mind when determining the nature and extent of the standardisation that may be necessary.

For almost all purposes identification elements have to be secure, unambiguous, efficient and manageable. The components of identification for example, name, sex, date of birth/graft and I.D. code, must be agreed and standardised. This is not to say however, that all components have to be common, there is some scope for reciprocal arrangements such as exist with EDTA and the Organ Exchange Organisation registries which incorporate each others' identification codes for mutual patient recognition.

Other items such as key data, special codes for static data, time-related variables and technical methods must all be considered. All seem sensible enough in the functional context but can be problematical when not generated from within the registry itself.

Numerous examples exist. Not all registries concerned with the HLA typing data in transplantation necessarily agree on the definition of specificity splits or of areas of cross-reactivity. The identification of patient sensitisation which depends on the type and size of a determining panel of test cells can present many obstacles to standardisation. This is true of any data where technical methods for its generation are not themselves fully standardised.

The relevance of the functional context and the care in evaluating the nature of standardisation must be constantly borne in mind in overcoming the difficulties presented. Quantitative ambiguity renders some data almost valueless from the statistical viewpoint but may be largely unimportant if used in a qualitative context. This is especially true where subjective data is involved.

A typical example of semi-quantitative data, presented as a numerical grade, being used to describe a subjective evaluation is that of rehabilitation status of patients. Some very complex and highly defined code structures exist. These require much work and evaluation on the part of the data provider if they are to be of value. Consequently such structures should only be used when the entity described is absolutely central to the study objectives. Where all that is required is a broad qualitative assessment of how well the patient is doing a sample three or four code structure is all that may be needed. A loose, broad analysis of a highly detailed code structure diminishes the purpose of the structure. Conversely a strict analysis of a loosely defined structure attempts to extract more information than exists or can be supported. In such situations as this, common sense decisions are necessary in determining standardisation.

Computerisation of data can add further dimensions to consideration of what to standardise, these are more likely to occur in the general context of how to achieve standardisation of the material data used by a registry.

How to Standardise

This subject can conveniently be dealt with under the two main headings of protocol agreement and data collection method, although it is important to be aware that these are neither independent nor mutually exclusive.

For a specified study or functional characteristic of a registry, the preferred method of standardisation would be that of protocol agreement. This approach ought to be fairly free of problems since participation can be largely secured by agreement (although, of course, the reverse is not necessarily the case).

As has already been suggested, data collection methods also offer possibilities for achieving standardisation. Perversely, manual methods seem to be best at achieving this in terms of their relative simplicity. All methods can be problematical however, for example, manual methods are prone to error and can be unreliable whereas computerised methods present compatibility problems. This is particularly true if the source of the data is computerised as well as the registry and the data is to be moved from one system to another.

The incompatibility between computer systems would seem to present a further case for more standardisation. Potential candidates for standardisation of data in this context are, at the lowest level, the stored form of the data (e.g. integer, decimal, character, radix and so on), next in the aggregation of data items into records and the accumulation of records into files. As discussed in the general context, it is necessary to consider what is practical and what is achievable without loss of value of the data.

The difficulties of physically transferring data held in one form on the data source system to the registry system where it will probably be stored in a different form are slowly diminishing. Generally applied standards exist for off-line methods of transfer using magnetic tape and cassette, but, far less so for discs. Also, the rapid evolution of networking technology and telecommunations between systems is very promising, particularly if the objectives set by OSI (Open Systems Interconnection) are eventually achieved.

Difficulties still remain however. No matter how easy it becomes to move data from one system to another, the differences in data types, record and file structures have to be resolved.

Imposition of rigid standardisation at any or all of these levels can be highly impractical, particularly if a registry has a wide catchment area for its data.

Alternatives

There are, of course, alternatives and it is highly appropriate that computers should be put to work to solve the problems that their makers create.

In simple systems it is perfectly feasible to develop specific structure translators. However, for a large registry, acquiring data from many different systems, the burden of developing many of these can be impractical.

Modern sophisticated data base systems, whilst simplifying matters of data storage and handling at the local and application level, do not necessarily improve matters unless, amongst the software tools available, a flexible formatter and assembler exists in the data donating system and its counterpart for the acquiring system is also available. At least, a defined transfer structure may be possible.

The present trend to produce standards for the data dictionaries that lie at the heart of data base systems and their associated data description languages will produce the means of understanding the structure of transferred data without the need to impose one. Nonetheless, the basic task of re-arranging

it for use in the registry data base still has to be performed. All too
often, where precise standardisation has proved difficult or impossible to
obtain, this remains the task of specifically written software.

A technique which can loosely be described as dictionary co-mapping is being
investigated for the EDTA Registry which, like CTS, has a very wide catchment
area for its data. A simple diagrammatic representation of dictionary
co-mapping is shown in Figure 3. This technique will be greatly simplified
when standard data dictionaries and data definition languages emerge. The
objective is to use the information embedded in these to map directly from one
system to another. In this way it ought to be possible to largely avoid the
compatibility problems without the need to standardise its structure.

It is appropriate finally to consider the advantages and disadvantages of
standardisation and how, in an imperfect world, to achieve the optimum
practical situation.

Advantages and Disadvantages

The advantages of standardisation may seem self-evident. The functional
efficiency of a registry's data handling is enhanced and, very importantly,
enables definitive analyses of the data. It could be argued that
standardisation will help, through comparability of data, to promote
collaboration and rationalisation of data gathering but, as has been
discussed, this may very easily turn out not to be the case.

All registries have compliance problems in one form or another. There are
many reasons for this but the imposition of standardisation can be one of
them. It is quite normal for collaborators and co-operating registries to
hold similar, but different forms of the same data of the type discussed under
the heading of what to standardise (HLA, biochemistry, coding sets etc). The
reasons for these differences can be very valid locally according to interest,
working practices and, indeed, opinion. Under such circumstances, rigid
standardisation can have far-reaching consequences for local workloads even
assuming the presence of the necessary expertise (e.g. in the individual
transplant centres). Once established however, long term standardisation can
help secure compliance through familiarity and confidence.

It should be evident from the brief consideration of protocol agreement and of
total standardisation for controlled, defined studies that there is little
scope for flexibility in such cases. Whilst there may be no need in certain
circumstances, it is important to be aware that the more complete the
standardisation then the less flexibility and ability to respond to changing
needs is available. It is very difficult to incorporate new aspects

mid-stream as it were, without compromising statistical validity, functional efficiency and compliance.

This is similar to and closely linked to limitations of scope that can result from standardisation, e.g. EDTA - main patient data is optimally suited for demographic studies but sub-optimal for other purposes. Clearly this has to be carefully considered in the functional design of a registry. Limitations of scope are acceptable if a registry's function, like a prospective study, carries a finite lifetime. They generally do not, of course, the reason being an unwillingness to abandon a functional infrastructure and attempts to use it to progresss to new areas of interest are commonplace. This brings additional problems in the form of a tendency to accumulate progressively increasing volumes of obsolete and irrelevant data.

In the practical world therefore, a transplant registry often needs to be able to evolve, progress and develop - all of the things, in fact, that make standardisation difficult or, conversely that standardisation makes difficult. Clearly a practical balance has to be achieved between the advantages and disadvantages on the one hand, and the functional value and validity of the work of the registry on the other hand.

Is it possible then to plan for and implement standardisation in such a way as to secure flexibility, opportunities to alter scope, without prejudicing compliance and statistical validity? With care, it ought to be possible.

Figure 4 illustrates five steps which should be taken in planning for standardisation. This is not sufficient in itself however and the question of possible evolutionary paths should ideally be addressed at this stage also.

Figure 5 itemises four practical headings under which these possibilities should be considered. Clearly it is not possible to foresee all eventualities but unless consideration is given to the most likely directions of change, a great deal of time and effort will be wasted in future restructuring to meet required progress.

The EDTA Registry illustrates some of these features in its linked layered structural design. The individual renal and transplanting centres form the core data. The core records link to the relevant patient questionnaire data sets and both can be linked to data sets relating to subjects of special interest studies.

To finish on an optimistic note, it has been the experience of the EDTA Registry, in some aspects of its work, that sensible standardisaion practices encourage and help to promote good disciplines in data generation and sampling without being restrictive or dictatorial. The balance can be struck but the striving to obtain and improve upon it is a continuous process.

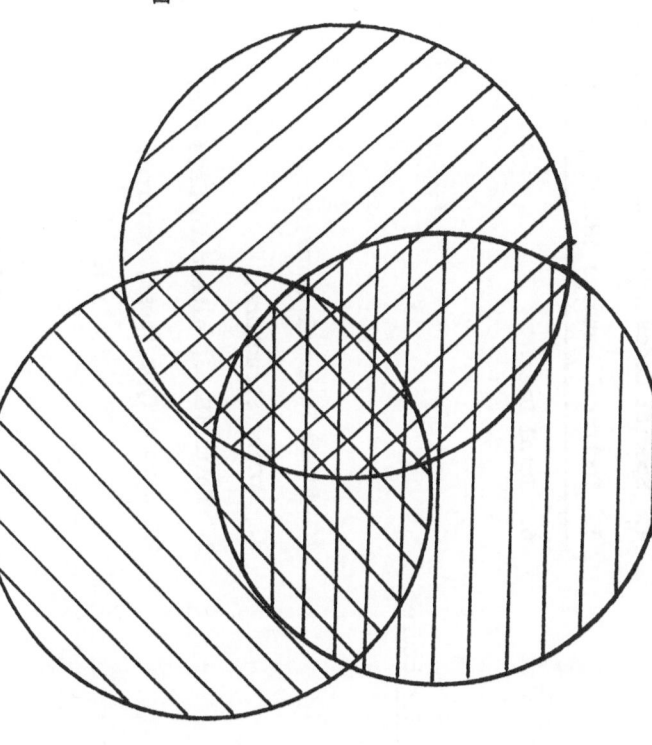

EDTA
DEMOGRAPHY
RESEARCH

CTS
RESEARCH

U.K. -, EURO -
FRANCE-TRANSPLANT
ETC

ORGAN EXCHANGE
RESEARCH
ADMINISTRATION

OVERLAPPING FUNCTION AND COMMON DATA REQUIREMENTS

Figure 1.

110

WHAT TO STANDARDISE

A. Data Content

1. Identification.

2. Key Data (ABO,HLA, sensitisation, dates etc)

3. Variables (biochemical data, drug doses etc).

4. Special Codes (PRD, CoD, CoF, drugs etc).

5. Technical Methods (R.I.A., Cytotoxicity, etc)

----

6. Total Data.

B. Data Structure

1. Fixed Interval Sampling.

2. Coding Structure.

3. Units (S.I., alternate days etc)

4. Sample Type (test, control)

Figure 2.

DICTIONARY CO-MAPPING

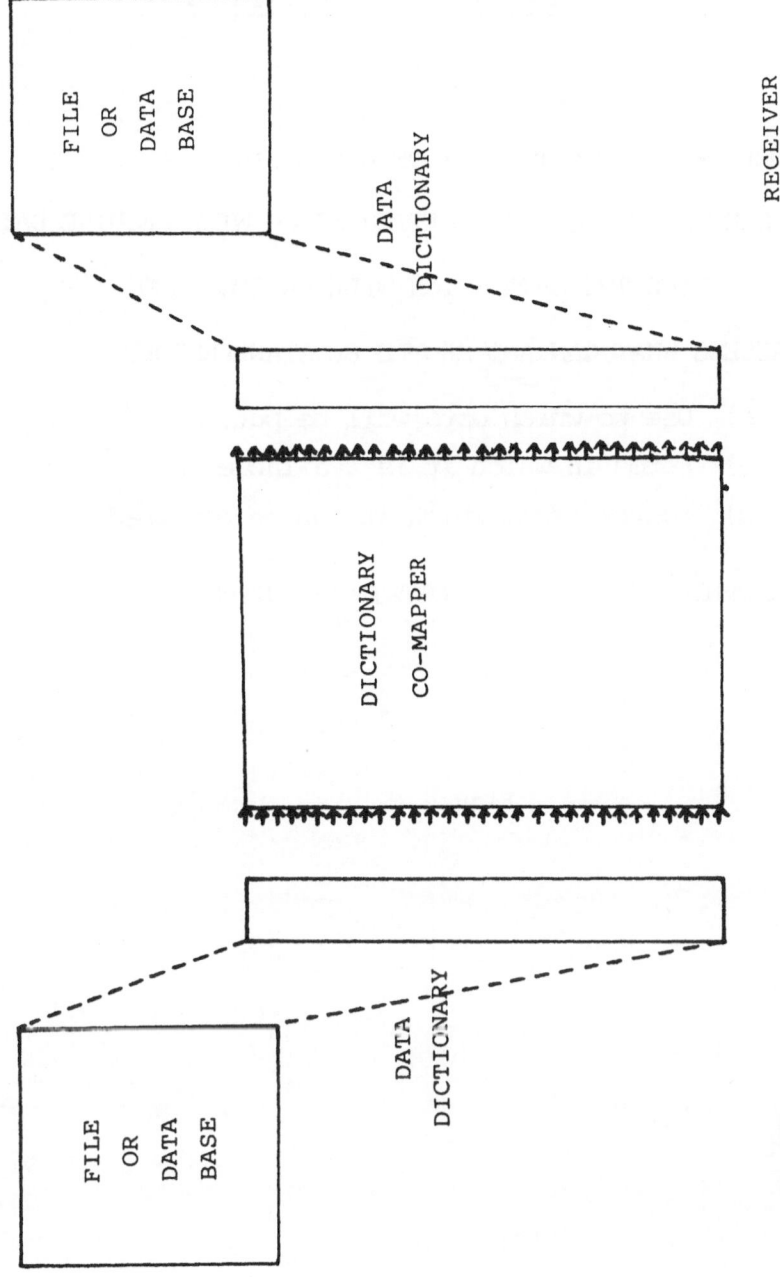

Figure 3.

PLANNING FOR STANDARDISATION

1. DEFINE THE FUNCTION(S) OF A REGISTRY.

2. IDENTIFY THE SUBJECT MATTER WHICH WILL REQUIRE DATA SUPPORT.

3. DETERMINE THE DATA WHICH WILL BE REQUIRED.

4. ASSESS WHAT CAN AND SHOULD BE STANDARDISED.

    a) Use to which data will be put.
    b) Forms in which it is available.
    c) Sources from which it can be obtained.

5. CONSIDER HOW BEST TO OBTAIN THE DATA.

Figure 4.

PLANNING FOR EVOLUTION

1. DEFINE A MINIMUM CORE DATA SET.

2. PROVIDE STABLE 'LINK' POINTS FOR PERIPHERAL STUDIES.

3. DETERMINE PATHWAYS FOR FUTURE EXTENSIONS OF DATA.

4. SELECT APPROPRIATE DATABASE SOFTWARE AND TOOLS.

Figure 5.

TRAINS

An Information System for Transplant Data

E. Keppel
IBM Heidelberg Scientific Center
Tiergartenstrasse 15
D 6900 Heidelberg, Germany

ABSTRACT: A transplant information system being developed in the framework of a joint research project between the IBM Heidelberg Scientific Center and the Department of Transplant Immunology of the University of Heidelberg will be presented. The system aims at exploiting computer network facilities to provide distributed information processing services to geographically disseminated transplant centers.

The system provides services on a central dedicated system to collect, manage, and evaluate relevant transplant data for various types of collaborative multi-center studies. At remote sites, PC front-ends are linked to the central transplant information system. These allow to perform remote database update of study data, make available some of the central services, and provide computer power to local users for their own needs. On the central computer specific tools, procedures, and reference data are accessible by remote transplant centers to support their daily clinical routine regarding immunological problems.

EARN is the broadcasting vehicle for information exchange. The system provides facilities for central study administrators to design databases for collaborative studies and to control central system access. Remote local autonomy and flexibility for adaptation to local needs is achieved by means of a distributed data dictionary which monitors the centrally controlled procedures.

INTRODUCTION

Over the past years organ grafts have become an increasingly used treatment for end stage organ diseases. Computer techniques have found a large field of application in transplant medicine as the demand of clinicians and researchers for the acquisition and analysis of large samples of transplant data serves the support and improvement of transplant techniques. There exist two broad areas of application for large computer based information systems:

- Operational systems, designed to support transplant centers in the day-to-day clinical routine. The services are provided at a central location, controlled by the common organization of a pool of users. The services include the maintenance of common lists of patients waiting for a transplant, the optimal allocation of available organs to potential recipients in the waiting lists, the evaluation of the degree of immunological match between donor and recipient, and the analysis of transplant results. Such systems focus on high availability and reliability of the system, coupled with powerful processing capabilities in order to satisfy the fast information retrieval needs of the transplant centers.

- Research systems, developed to cope with long term medical research and based on the evaluation of large samples of transplant data. They aim at supporting the establishment of common transplantation strategies and care methods. Flexibility and adaptability to rapidly changing needs, ability to interface with evaluation software tools, and efficiency in processing large amounts of data are the key requirements put upon research systems.

Common to both types of systems is to operate in an environment of collaborating transplant centers, widely disseminated geographically, and far away from the central location where the information is collected and processed. Data submission is performed by slow, unreliable and cumbersome manual procedures like filled-in forms, paper listings, phone calls, etc... Information retrieval by endusers, if at all available, requires equally long time delays and expenditure of bureaucratic procedures. It is generally recognized that the most serious shortcomings of existing information systems for processing transplant data are caused by the lack of efficient, reliable, and rationalized communication means between the central site and its remote users.

This situation is based on the fact that traditional database management systems (DBMS) are centralized systems which do not support long range data communication and remote information processing in a way that adequately serves decentralized information processing, i.e. able to exploit the autonomous powerful processing capabilities offered by the PC workstations available today on the market. In distributed systems, PC's are exclusively used as remote terminals to access the resources of a central computer system Although for the past years a great deal of research effort has been put into the investigation of principles and methodologies of distributed processing, no general purpose commercial software is offered which meets the needs of decentralized operations. Existing systems are either tuned to serve very specialized applications (e.g., air reservation), or are research prototypes in experimental stage.

The subject of the present paper is to present the prototype implementation of a distributed information system for transplant data, able to manage the cooperation of remote PC workstations with a central information system. The paper is structured in four parts:

Section 1 presents the framework of the project and describes the requirements put upon the system, aiming at supporting multi-center evaluations in organ transplant medicine.

Section 2 addresses the concepts and methods to implement the projected system. It focuses on some of the critical issues applicable to the prototype.

As the central body of the paper, Section 3 describes the prototype implementation of TRAINS. This section highlights the general architecture and sums up the major services offered to the users. Implementation and some coding techniques are briefly addressed.

To close the paper, Section 4 will report the project status and give an outlook of the planned activities.

1. PROBLEM STATEMENT

1.1 Project Origin

The international Collaborative Transplant Study (CTS), run for several years to evaluate worldwide kidney transplant data, is the initial motivation for the research project presented in this paper. Since 1982, patient histories are collected by Prof. G. Opelz at the Transplant Immunology Department of the Heidelberg University and permanently evaluated to gain global insight in immunological processes and transplant patient care. A database system was set up very early in Heidelberg to transfer the received patient data into the computer and to perform multi-center evaluations and analyses [Hen83]. The results are periodically returned to the participating centers in form of paper listings (patient lists, follow-up questionnaires, etc...) and graphics (survival curves). With the growing success of the study the number of participating centers increased rapidly, causing a fast increase of the manual workload at the central location. By 1984, the database system was no longer able to process efficiently the large amount of data [Koe86]. Today, appr. 300 transplant centers submit data to the CTS study, contributing to a central data collection of more than 40,000 patient histories.

The Heidelberg Scientific Center, a research unit of IBM Germany for advanced software technology, has been attracted by the CTS application as an exemplification of distributed information processing. A joint research project has been agreed upon with the University of Heidelberg to gain knowledge about the requirements of computer networking and interactive problem solving environments in the medical environment. It has been agreed to develop commonly the prototype software of a transplant information system integrating wide area networking facilities to achieve distribution of processing.

1.2 Project Objectives

To get insight in the key question, to what extent the decentralization of processing may improve the efficiency of collaborative studies, an overall distributed concept has been defined. It is illustrated by Figure 1:

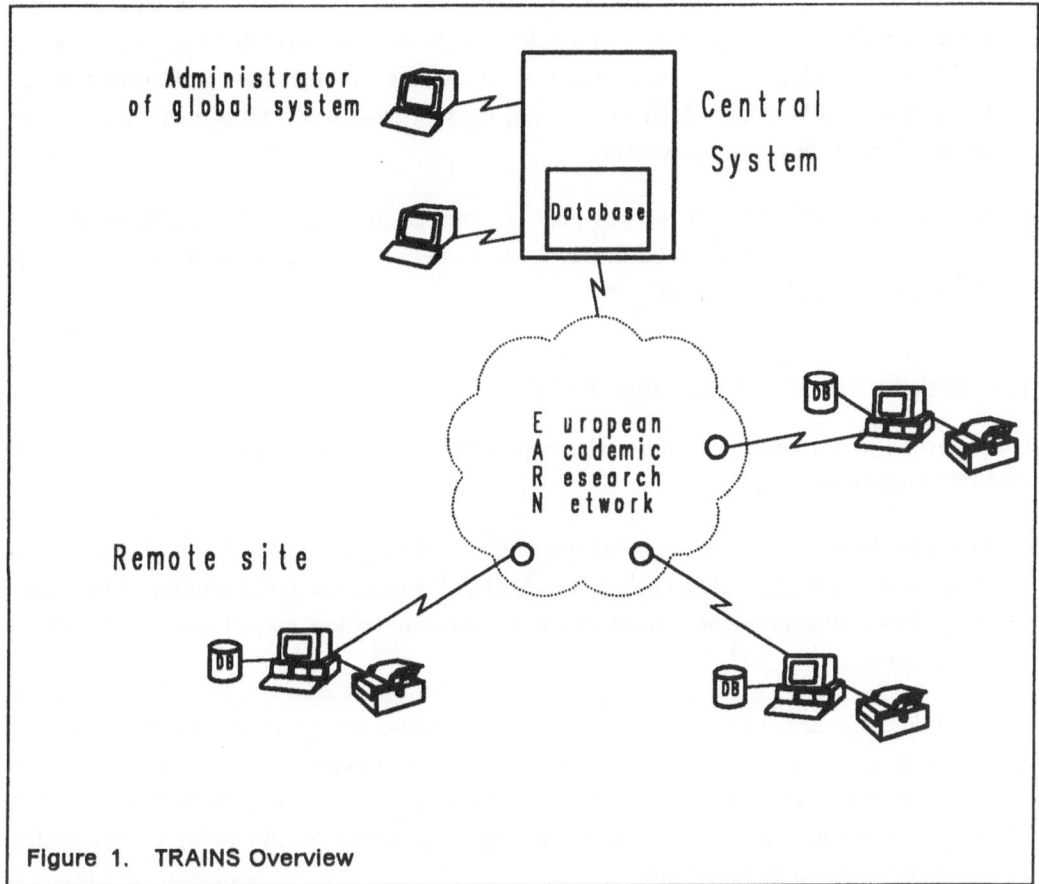

Figure 1.   TRAINS Overview

A central main computer, dedicated to the project, is set up at the Department of Transplant Immunology of the University of Heidelberg to control the overall information system. A database management system (DBMS) manages the central database, repository of the relevant common data of collaborative studies. It serves for large scale evaluations and statistical modelling on the total amount of data.

To collect the data and to drive an active cooperation between the central site and the transplant centers, PC based workstations are made available at the remote locations. A data communication link is established with the central site by means of the computer network EARN, which will be described later. Objective of the project is to develop the software for achieving the control of cooperation for the total system.

For collaborative studies, like CTS, the following major benefits are expected:

- Data entry is off-loaded to the transplant center, close to the location where the data originates. Errors may be immediately corrected. As the overall validity of data input is better controlled, a relevant improvement of the data quality is expected

- The transplant centers obtain the responsibility for their own data. A local copy of the center's study data is kept as a local database of its own patients. By this means, services which were available uniquely at the central site are permanently implemented on the autonomous workstation: Listings, queries, reports, local evaluations, can be performed independent of the central processor.

- Some of the central system services may be accessed at any time by remote users. For example, standard central evaluation tools, or global reference data, permanently updated at the central site, become available.

1.3 Requirements and Usage Policies

The following design guidelines summarize requirements and usage policies established for the TRAINS prototype:

- *Centralized study control:* The central site defines and manages exclusively the common data of collaborative studies. The central site also has the exclusive access control of the global study data available in the central database. Remote centers cannot access each other's patient data.

- *Flexibility:* No dedication to the CTS study operations should exist. Study parameters should be easily adaptable to collaborative studies and applications of similar structure. Flexibility for modifications and gradual extensions is a key requirement for an easy administration of study data. Moreover, study operations as run today on the central system will have to be supported.

- *Extendability:* A modular design should guarantee smooth integration of new local or distributed services and applications. The areas of interest are:

  - The maintenance of patient waiting lists common to the collaborating transplant centers

  - Central decision support for optimal donor/recipient match and organ allocation

  - Registration and analysis of common special patient groups

- *Statistics:* Specialized tools for fast statistical evaluation of large amount of data are to be provided, as well as simple interfacing to standard evaluation tools (studies on special patient sub-populations)

- *Ease of use:* Usability by non-DP professionals is a key issue to focus on, in order to allow people in the clinical routine to easily handle the system. The enduser services should be provided by means of uniform and easy-to-learn, dialog-driven interfaces.

1.4 The European Academic Research Network (EARN)

The broadcasting medium for the projected information system is the European Academic Research Network (EARN) [Heb85]. EARN is a wide area computer network which links by means of leased telephone lines, approximately 700 mainframe computers servicing most of the European universities and research institutions. It enables the exchange of data files by means of standard basic IBM communication products. However, EARN is an "open" network: As most of the computer manufacturers provide communication software which emulate IBM standards, the network includes a variety of non-IBM mainframe computers (DEC, SIEMENS, CYBER, etc...)

Gateways also allow communicattion with similar networks established in the USA, Canada, Japan, and Australia [Zor85]. The total number of mainframe computers linked together is approximately 1,900. The total number of computer users able to communicate with each other over the world via the network is estimated to surpass 50,000.

For the TRAINS project, the central computer is included in EARN as a node. Unfortunately, the PC based workstations projected as remote front ends to the central system can *not* be nodes of EARN (There are several technical reasons for this. The most prohibitive is the lack of adequate basic communication software for PC.) To build up a data communication link with the central site, the PC workstation must be attached first to a close host computer, node of EARN. Communication is made via a server machine defined in the host computer. TRAINS has to provide a software to drive PC-Host data upload-download and the communication process in the server. More details will be given later.

2. CONCEPTS AND METHODS

Expressed in terms of computer science, the objectives of the targeted information system entail the solution of specific issues linked with the development of *distributed* applications. The technical challenge of developing a software to control the cooperation of several, physically dispersed autonomous computer systems is an extremely complex and highly challenging issue, for which no complete solution is expected in the near future. The major aspects of distribution which make our system easier to implement are (1) the projected centralized control and (2) the restrictions regarding data access from remote sites, as defined in the initial usage policy statements. In this section, we will discuss the major issue areas applicable to the development of our prototype.

2.1 Problems of Distributed Information Processing

The obvious major objective of a distributed application is to make transparent to the users, as much as possible, the dispersion of processing on multiple computers and geographic sites. This is achieved by a set of appropriate techniques and methods, such as data replication on remote computers and specific communication control mechanisms (protocols) which tend to provide a unique logical view of the global application [Cer84]. Ideally, a distributed information system should look like running on a single central computer. Applied to the projected information system, the following technical issues of distribution need to be assessed and specific practical solutions provided.

- **Communication.** When using EARN to transfer data, *no permanent link* is established between the sending and the receiving site. The usual end-to-end protocols which are applied to communicate from a remote terminals to the resources of a central computer system are not applicable. Additionally, the instability of file transfer speed and network partitionings due to inactive nodes may lead to extremely high data transfer delays. Special communication protocols are to be developed to control communication.

- **Software/Hardware Heterogeneity:** PC workstations are expected to provide the local computing power at the remote site. Their integration in a centralized distributed system controlled by a software developed and running on a mainframe computer presents a number of typical problems due to the high degree of heterogeneity of software and hardware [Reu87a]. The major inconveniences can be summarized as follows:

  - PC-Mainframe communication can be performed, today, only by means of slow and non-flexible download-upload facilities. Moreover, no standardization exists between different manufacturers. TRAINS will have to deal with restrictions and inconsistencies regarding the PC-mainframe link software available at the EARN node of the connected transplant center.

  - Replication of global system information on the workstation is difficult to control from the remote central computer. PC based database management systems differ from mainframe systems.

  - Mainframe and PC operating systems are incompatible (different data encoding, access methods, etc...)

- **Data fragmentation** concerns the optimal splitting and allocation of data onto the different computers of the global system in order to implement the most efficient data access of the local applications. The objective of fragmentation strategies is to achieve efficient data replication schemes and update management.

The clinical data collected in the transplant information system is partitioned in disjoint subsets: single patient data, encompassed to the patients of a transplant center. Mutual dependencies between neither the single patients data records nor between the different centers data are to be enforced (apart a clinic code an a unique patient identification key). Therefore, an obvious and simple fragmentation scheme can be adopted: a partial redundancy of the total data collected in the central database, and consisting of the local copy of the clinic's own data, will be maintained by the global system. This replication organization greatly simplifies transaction management and data integrity control, as discussed in the next sections below.

- **Transaction management and recovery:** Transactions are the basic processes of any information system to transfer data from one location to another such that the system wide consistency of the data remains intact at any location [Dat85]. For example, when a patient update is issued at a remote workstation, the update of the local database can be committed only when a copy of the data has been received at the central computer and acknowledged to the sender. Otherwise an "undo" of the entire update must performed.

Transactions consist of a set of elementary operations made up of data transfers, acknowledgments, or rejection statements, and database manipulation operations. In the case of our distributed system, specific transaction and recovery procedures are to be developed to coordinate the consistency of central and local database. The issues are:

- No permanent communication link is maintained over the network between sender and receiver, leading to specific complex commit protocols to be developed

- Timeout techniques to detect system crashes and network partitions are difficult to handle

- Communication can be very slow, prohibiting database locking to manage database transactions

- For recovery, database undo procedures, consisting of restoring patient data from a remote primary copy, require complex mechanisms

- Transaction management must be implemented on heterogeneous hardware and software.

To overcome these problems, specific control mechanisms have been developed in TRAINS. As implementation details are outside the scope of this paper, no deeper insight in transaction management will be presented.

- **Data integrity** encompasses all aspects of processing which impact the preservation of correctness and consistency of the data collected and manipulated. Many data manipulations may by conceived which lead to the violation of the data integrity of the information system. Some integrity constraints are enforced by the used DBMS, most of them must coded in the application programs.

For the transplant information system being designed, the following integrity constraints are relevant and must be emphasized accordingly:

1. Data entry validation

   - The rules to assert the validity of individual data go beyond the classical assertions integrated in traditional DBMS (data type, relationships, etc...). Validity control requires the development of complex assertions which can be stated only with a high level programming language.
   - Mutual dependencies of the data attributes within a patient record.
   - Automatic and reliable generation of patient identification keys.

2. Operational integrity

   - A strong reglementation of authorized database manipulations must be required. In collaborative studies, transplant centers commit to provide periodically patient data which underlay strongly predefined standards and conventions. Therefore, database editing transactions to insert, update, or delete patient data must be strongly reglemented.
   - The system should ensure that data of individual patients is complete and consistent over the whole database (referential integrity).

2.2 Managing the Global System

Another aspect of the problems addressed by the present prototype is the consideration of how the features of a traditional centralized control of database management system appear in a distributed environment. The complexity of overall control obviously increases as more people are assigned to monitor portions of the global system. In this section, the aspects addressing structure and flexibility of the global system management will be discussed.

2.2.1 System Administration

The centralized responsibilities of system administration are structured less strongly than in a traditional central system since remote computers operate as autonomous workstations. Moreover, at the central site, responsibilities for controlling individual studies have to be distributed among the personnel supporting the global operations. Therefore, a hierarchical structure of administration is necessary, with a gradation of access authorizations to the global database resources in order to avoid system inconsistencies and side effects:

1. At the central site, a **TRAINS Administrator** is responsible for overall control of the central computer resources and the global DBMS available for all studies.

2. For every study to be run, a central **STUDY Administrator** is appointed to design and control the portion of the central database allocated to the study.

3. At every remote center, the single user workstation is provided with some administration functions in order to define and maintain local studies, similar to centralized studies.

As administration encompasses multiple and complex activities, to be shared among several users, care must be taken to provide the functions with a high degree of flexibility and user-friendliness.

2.2.2 The Data Dictionary System

To achieve the high degree of flexibility and usability required for study administration a comprehensive Data Dictionary System (DDS) has been developed as fundamental part of the system.

A DDS is a software tool used to describe, store, and maintain the total information available at a given time in the information system [All82, Cur81, Ref86]. It consists of a set of internal data files (Catalog) which describe the application data and their relationships, and may include any information regarding data descriptions, procedures, programs, user interfaces, directories, access paths, etc... The Catalog is itself a database, containing data about the application data. The design and management of a DDS is an intricate task but multiple benefits can be gained from a comprehensive system:

*   It allows the administrator to easily design and modify a study.

*   It is the most complete and up-to-date documentation of the study data and operational processes.

*   It achieves software and hardware independence since data is accessed without knowledge of the storage characteristics of the underlaying DBMS. Migration of the study schema onto workstations is easily performed as the DDS may be mapped on any hardware and software as long as the same data model is used.

*   The dynamic access of application programs on the Data Dictionary provides a purely *data driven* processing. Studies are designed and modified by means of interactive data editing, without changing program code. Therefore, for extensions, the high costs of skilled programmers for software development are mainly reduced.

The structure of the TRAINS catalog is given in Appendix.

3. TRAINS DESCRIPTION

3.1 System Overview

The implementation of the TRAINS prototype was started in 1985. The facilities have been implemented using relational database management systems: SQL/DS for the central site, and dBase III for the remote front-end to the system. This section is devoted to an overall description of the system, its global architecture and a summary of the functional capabilities available today.

Figure 2 displays the logical structure of the prototype.

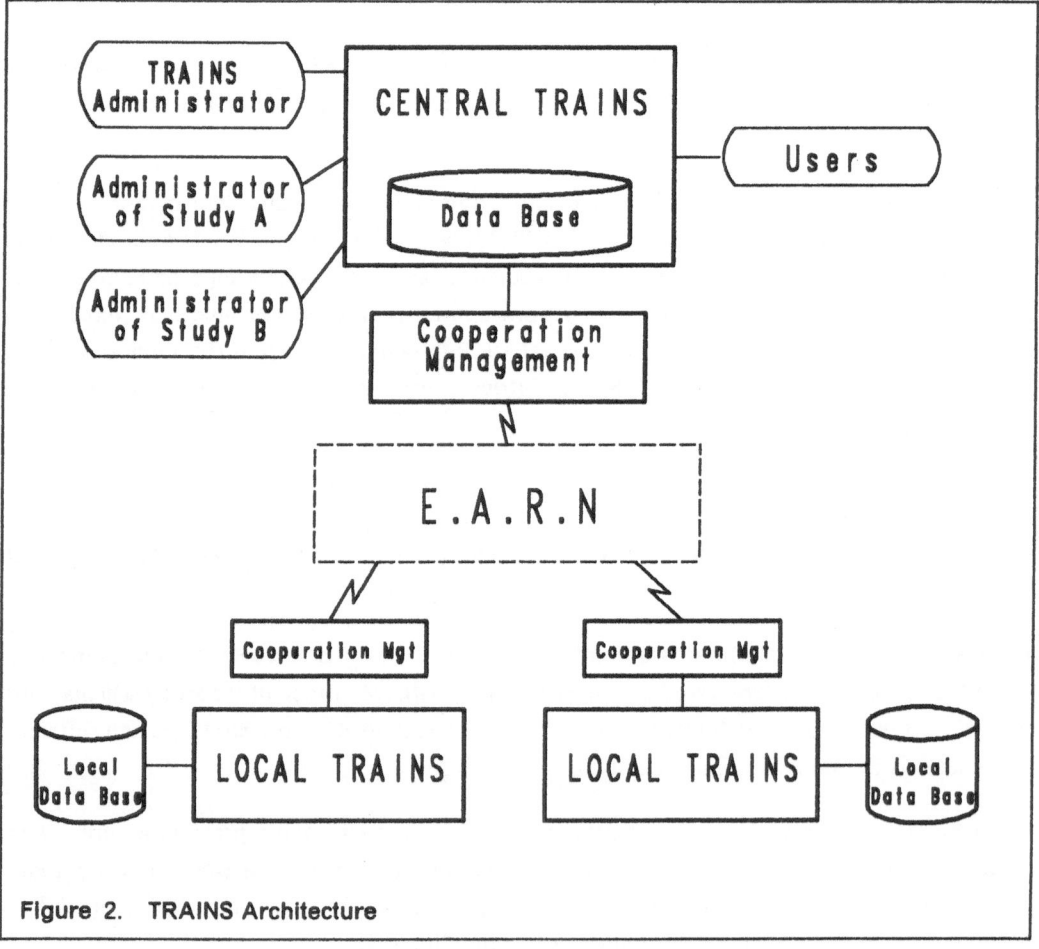

Figure 2.   TRAINS Architecture

The following three components constitute the total system:

1. **A central component**, controls the overall operation of the system. The software serves to support:

   - Administrators who set up and monitor either collaborative multi-center studies, or purely local applications
   - Personnel who perform mass data entry of transplant data transmitted via paper forms
   - Central users who run evaluations and perform statistical modelling on the studies data.

   The detailed summary of services available at the central computer will be given in the next section.

2. **Local components**, based on workstations, to provide autonomous computer resources to transplant centers. The software allows data entry of the patient data under control of a local copy of the central data dictionary. Entered data is collected in a local relational database, and immediately transferred to the central database. Evaluation tools are provided for local evaluation and for setting up local studies, using the services of the front end software.

3. The **Cooperation component** interfaces the distributed components of TRAINS with the communication network EARN. Its function will be presented later.

3.2 Central System Services

The logical structure of the central component is shown in Figure 3. The data dictionary is the fundamental component of the system, as it controls all system facilities: It provides a purely data-driven access to the central database, it is the base for the dialog management, and it controls the cooperation with the remote sites. A detailed description of the data dictionary structure is given in the Appendix.

The following functional capabilities are available, provided by means of easy to use, dialog-driven, facilities:

1. Administration

System administration is shared among several central users. The administration hierarchy is as follows.

1. A central TRAINS administrator. His tasks include:

   - Global database creation and management (backup, recovery, privacy, security)
   - Allocation of Database resources to the respective studies
   - Identification and enrollment of study administrators, see below
   - Link control of remote transplant centers (maintenance of a network directory)
   - Management of the global TRAINS catalog

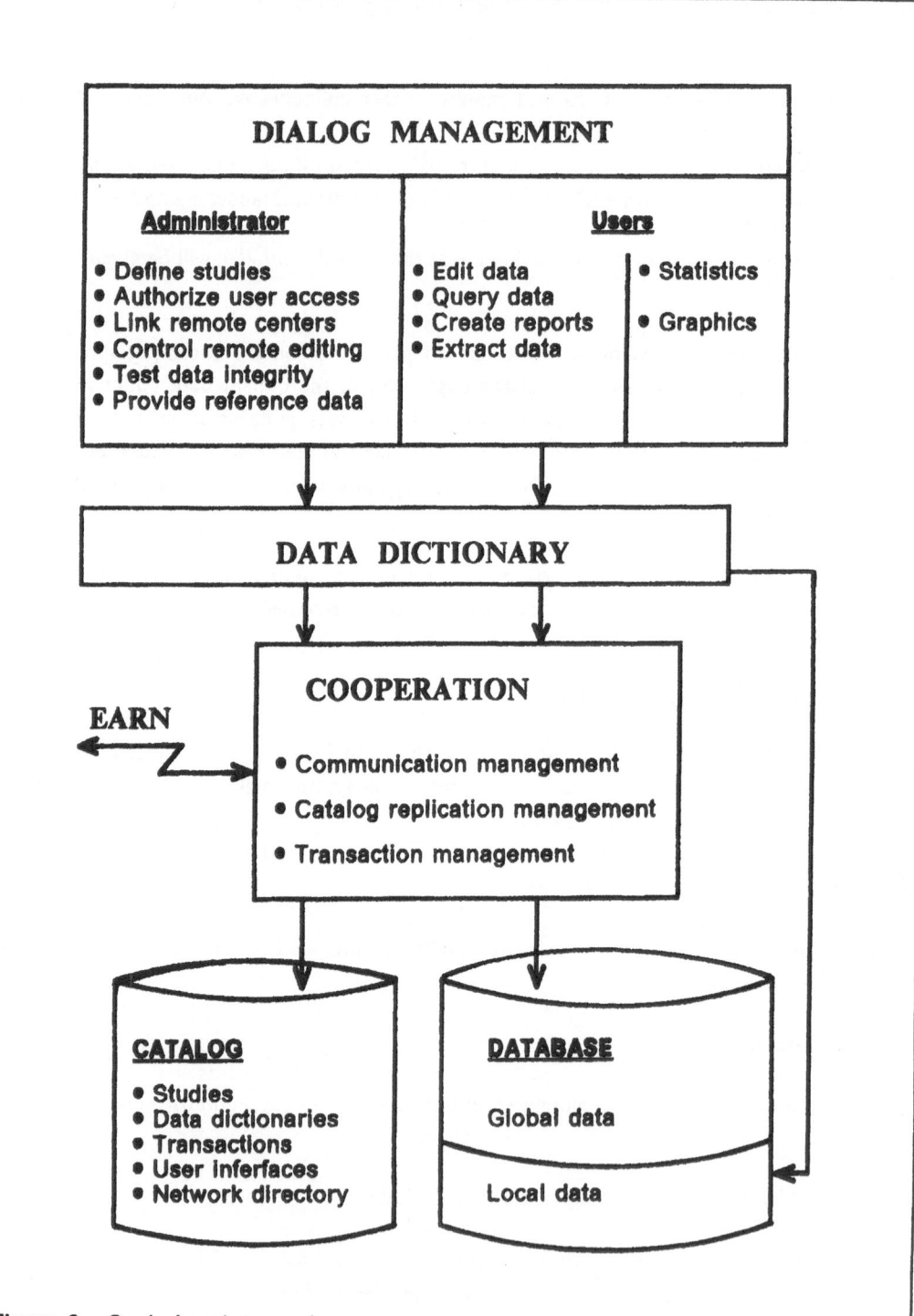

Figure 3. Central system services

2. Study administrators, responsible for:

   a. Definition and maintenance of a study, i.e.:

- Database definition of study data (i.e., structuring the study data entities and attributes in tables and views) and definition of integrity constraints (validation rules and consistency correlations)

- Data manipulation reglementation, i.e., the design of database editing transactions and user interface panels

- Definition of user interfaces for on-line queries, data extractions, and batch reporting.

   b. Operational administration:

- Control of central and remote user access to the study data

- Control of central editing of remote data (data transmitted via filled-in forms to the central site)

- Control of overall database integrity

- Monitoring of the central evaluations and reporting (creation and sending out of patient lists, follow-ups, evaluations)

- Extractions of sub-populations data and statistical evaluations

- Management of database schema changes, including update dissemination to remote centers

- Preparation of global transplant reference data for remote retrieval

Study definition and maintenance is implemented as interactive data dictionary manipulation, using the the database editing facilities provided by the system (see appendix) This procedure provides an iterative approach to logical database design for setting up a study. Database design includes the generation of test data, panel design, and facilities for on-line testing of editing transactions.

2. General Users

The following facilities are available for central database editing and analysis:

- Data editing: input, update, or deletion of patient data.

- On-line queries and reports

- Running of predefined batch reports

- Extraction of data subsets

- Interactive statistical analysis. A package has been developed to process efficiently large amounts of data [Reu87b]. Sub-populations may be periodically extracted, the data reformatted to inverted files and survival curves analyses performed in a highly interactive way.

3.3 Remote Systems Services

Figure 4 displays the logical structure of the services available on the PC workstation at remote centers. As one can see, it is a single user subset of the central system. Again, the data dictionary is the heart of the system. A copy of the central catalog version is transferred as the center is linked to the central site. Catalog update is controlled by the communication component. The dictionary provides the study data definition, the legal editing transactions, and the user interface panels. The dictionary also allows mapping of the central study database onto a local relational database management system. The data dictionary may also be used to define study data for local usage.

The software provided on the remote workstation provides the following services for centrally controlled studies:

- Data editing, i.e. insert, update, and delete study data

- On-line query and reporting of the local database

- Local evaluation tools by means of data extracts

- Remote access to central reference data

3.4 Cooperation Management

The cooperation management is the software to provide a reliable and consistent communication between transplant centers and the central computer. It is mainly responsible for three tasks:

1. Data transfer from the sender to the receiver site, ensuring the synchronization of the broadcasted files (data and messages). It also assures encryption and decryption of the data on the public transmission lines.

2. Transaction management and recovery. It is achieved by two distinct mechanisms:

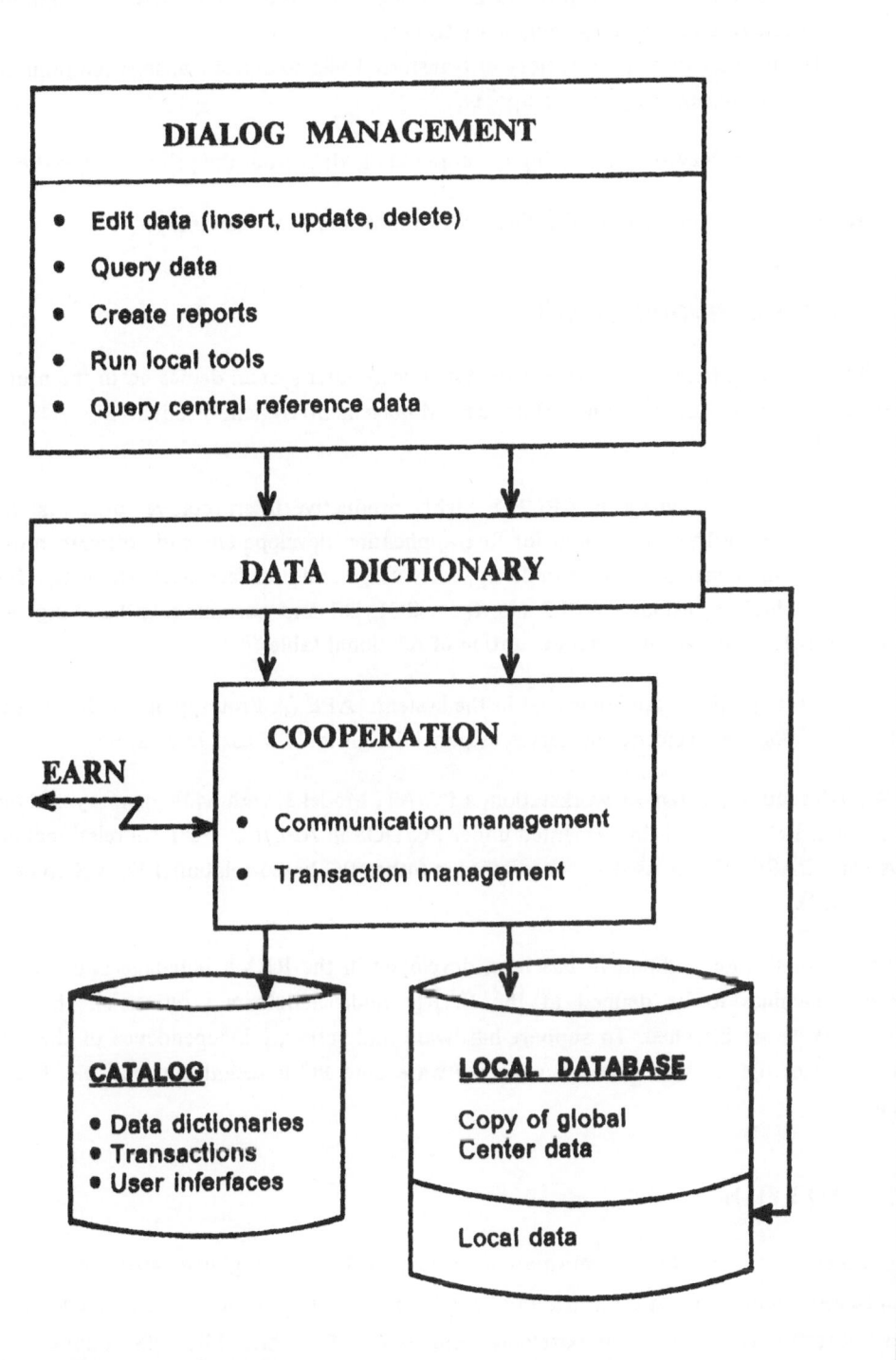

Figure 4.   Remote system services

a. A Two-Phase commit procedure, developed on the top of EARN communication protocols to control long range data transfer

b. The maintenance of a directory of transferred files to detect transmission failures and to control concurrent database update.

Recovery is achieved by reloading the entire clinic data from the primary copy site.

3. Replication management of the TRAINS catalog and data dictionary.

3.5 Implementation Details

TRAINS is being implemented on an IBM 4381 computer system dedicated to the project. It is developed around the relational Database Management System Product SQL/DS, running under VM/SP.

The programming language is APL2, a highly productive interpretative functional language which is particularly well adapted for fast application development and software prototyping techniques: incremental modular build up, early usage and user feed back, stepwise refinement, ease of gradual extensions. APL2 interfaces SQL/DS and permits easy handling of nested arrays which are provided as representation of relational tables.

IBM program products are integrated in the system. APE (A Prototyping Environment) is the base for dialog management and serves as panel design tool for user interfaces.

TRAINS requires, as remote workstation, a PC/AT, Model 3, with 640K memory and one fixed disk. The PC software is implemented under PC/DOS in APL/PC. As local relational database system, dBASE III + is used in the prototype (other PC based relational DBMS could be also integrated).

The communication component has been developed in the REXX language and runs on VM server machines to be defined at the EARN node mainframes on which the TRAINS workstations are attached. To support hardware and software independence of the system, a migration of the REXX communication software onto PC is technically possible, but not yet made.

4. OUTLOOK

Implementing a distributed information system such that a global software manages an application running on an arbitrary number of dispersed autonomous computers and controlling several remote databases is an extremely complex task for which TRAINS definitely does not provide a solution. The present system could be successfully implemented since the requirements of collaborative studies present a number of simplifications, which can be exploited to reduce the software complexity:

- The centralized control of overall operations and the simple data fragmentation. Intricate data replication management and multiple update control can be neglected.

- The prototype does not handle the difficult problem of managing a distributed database

- User-controlled procedures allow bypassing the complexity of automated transaction management and recovery procedures

The contribution of TRAINS to distributed information processing in transplant medicine should be highlighted:

- The functional flexibility is demonstrated by the successful operational usage of the central services at the University of Heidelberg. Several international collaborative studies have been running since 1986 (kidney transplant, heart transplant, highly sensitized patients).

- Flexibility to changes: Interactive catalog management allows easy creation and modification of study databases. Modifications are rapidly integrated in the operational system since no change in program code is required.

- A relevant improvement of data quality has been achieved by means of the large number of complex integrity constraints defined on the study data.

The PC software for remote data entry on workstations has been completed and put in operation at selected pilot sites. Recently, the first CTS data was input over EARN to the central system. It is planned to increase gradually the number of transplant centers linked to the system and start the evaluation of the distributed capabilities of TRAINS.

APPENDIX: The TRAINS Data Dictionary System

The TRAINS Data Dictionary System (DDS) is build upon a catalog of internal SQL/DS tables stored in the database of the central system and controlled by the TRAINS software. It contains the relevant information available at a given time on all studies and processes. It monitors system and study administration, cooperation control, transaction management, and recovery.

System administration by means of the DDS is implemented by interactive manipulation of the catalog data, using the database editing facilities provided by TRAINS for endusers. Predefined catalog update transactions with extended built-in integrity constraints achieve a strong control of the global consistency of the system catalog.

To indicate the amount and type of information managed via the DDS, the table below lists the TRAINS catalog. It displays the internal table name, a description of its content, and the number of columns, which is an indication of the amount of information controlled. Information is structured as follows:

TABLE NAME	CONTENT	Number of Attributes
S_APLIC	Description of Studies	27
S_AUTHV	User access authorizations	5
S_APTAB	Tables and view	11
S_DDESC	Data descriptions	21
S_UINTF	User interfaces and transactions	11
S_NETWK	Directory of linked centers	8
S_SCTAB	Online query scripts	15
S_REPRS	Batch reports	10
S_CFDIR	Directory of Transfer Control Files	39
S_MLOG	Transaction management control	7

Table I: Overview of TRAINS Catalog

The following information is contained in the tables. Some information is controlled by the administrators by means of the built-in data editing facility of TRAINS. Most of the information is, however, managed by the system software itself.

S_APLIC   Defining a study consists of creating an entry in this table. Following information is maintained:

- A unique study code and a free text description
- Study administrator
- Network identification of sender (set by system)

- Network identification of receiver (set by system)
- Database schema identification
- Link status

S_AUTHV   Granting a user to access the facilities of TRAINS creates an entry in this table.

S_APTAB   An entry in the table is created for any administrator request to create an SQL Table or View. Following information is maintained:

- Internal identification, study code, and a long description
- A TRAINS attribute code (local or global table)
- SQL/DS information: Name, creator, DBSpace,
- Status
- Change control identification

S_DDESC   An entry in this table is created for every attribute (data field) of a study. This table is a total description of all data attributes available in the global SQL database, and controlled by TRAINS. Following information is maintained:

- Internal identification, study code, and a long description
- Data description: Name, type, etc..
- Access information to SQL/DS
- Integrity constraints

S_UINTF   An entry is made for every editing transaction or user interface (browse, on-line query, extract). Following information is maintained for every data field:

- Study code and transaction long description
- Transaction type (local, global, test mode, inactive)
- The S_DDESC entries of the fields to be manipulated are to be individually referred to, adding following information:
  - Field edit attribute (input, output, or calculated field)
  - panel name
- For insert transactions, a key generation type (sequential, or a specific TRAINS patient identification key)

S_NETWK   This table contains the EARN identification of linked remote centers and a link status managed via TRAINS

S_SCTAB   This table contains saved dialog scripts to request a database on-line query (SELECT statements on the database), or a database extract

S_REPRS   An entry in the table is created for every batch report defined in the study

S_CFDIR   This table is an internal directory of all data and messages files, named Transfer Control Files (TCF), making up a transaction. This table is exclusively controlled by the transaction and recovery manager

S_MLOG   This table is an internal log file for transaction management and recovery.

Modifications are performed as editing transactions of the dictionary database. For this, the editing facilities provided to endusers for the manipulation of study data are available to the study administrator for dictionary maintenance.

References

All82   Allen, F.W., Lomis, M.E.S., Mannino, M.V., The integrated Dictionary/Directory System, Computing Surveys, Vol.14, No. 2, p. 245-286, June 1982

Cer84   Ceri, S., Pelagatti, G., Distributed Databases, Principles and Systems, McGraw-Hill, Ney York, 1984

Cur84   Curtiss, R.M., Little, A.D., Data Dictionaries, as Assessment of Current Practice and Problems, Proc. of the 7th Int. Conference on Very Large Databases, 1981

Dat85   Date, C.J., An Introduction to Database Systems, Fourth Edition, Addison-Wesley Publishing Company, 1985

Heb85   Hebgen, M., EARN - Ein Computernetzwerk für Wissenschaft und Forschung in Europa, Das Rechenzentrum, Vol. 8, Nr.1, 1985

Hen85   Hennige, M., Implementation of an International Information System on Kidney Transplantation, Methods of Information in Medicine, 24, p. 135-140, F.K. Schattauer Verlag GmbH, 1985

Koe86   Koehler, C.O., Engelmann, A., Keppel, E., Opelz, G., Hennige, M., CTS, Ein Internalionales Informationssystem für Nierentransplantationen, in: C.O. Ehlers and H. Beland (Eds.), Lecture Notes on Medical Informatics, Proceedings of the GMDS 31th Annuary Conference, Göttingen, September 86, Springer-Verlag, 1986

Mar83   Marti, R.W., Integrating Database and Program Descriptions using an ER Data Dictionary, in: C.A. Zehnder (Ed.), Database Techniques for Professional Workstations, Institut für Informatik, ETH Zentrum, CH 8092 Zurich, Switzerland, 1983

Reu87a   Reuter, A., Haberhauser, F., Peinl, P., Zeller, H., Weber, D., Speicher, A., Friedlein, K., Renschler, J., Anforderungen an ein arbeitsplatzorientiertes Datenhaltungssystem, in: H.J. Schek, G. Schlageter (Eds.), Datenbanksysteme in Büro, Technik und Wissenschaft, GI-Fachtagung, Darmstadt, April 1987, Springer-Verlag, 1987

Reu87b   Reuter, R., Janßen, R., Interactive Analysis on Large Databases, Ebenda

Ref86   Reference Model for DBMS Standardization, Database Architecture Framework Task Group (DAFTG) of the ANSI-X3-SPARC Database System Study Group, SIGMOD RECORD, Vol.15, No 1, March 1986

Zor85   Zorn, W., Rotert, M., Lazarof, M., Zugang zu Internationalen Netzten, in: D. Heger, G. Krüger, O. Spaniol, W. Zorn (Eds.), Proc. GI/NTG-Fachtagung, Kommunikation in Verteilten Systemen, Karlsruhe 1985, Vol. 2, Springer-Verlag, 1985

DEUTSCHES FORSCHUNGSNETZ (DFN)-

DATA COMMUNICATION SERVICES FOR THE SCIENTIFIC COMMUNITY IN GERMANY

K. Ullmann
DFN-Verein
Pariser Str. 44
D - 1000 Berlin 15

1. Introduction

The functionality of DFN is defined by a set of data communication services and an underlying architecture, which provides these services to the user. Both items (service funcionality and architecture) will be described in chapter 2. Implementations of systems like DFN are normally based on clearly identifiable engineering principles. The three most important ones (role of standards, industry oriented quality measures for the software production and the role of manufacturers) are motivated and outlined in chapter 3. Chapter 4 contains four concrete application scenarios which serve as examples for the use of DFN.

Chapter 5 gives an outlook to some future developments within DFN. It is based on a review of the application scenarios and motivates some of the future topics of DFN-developments.

2. DFN-Services and their Architecture

2.1 DFN Services

DFN-services are developed for the scientific community. They can be used by other users as well - there is no principal technical or administrative reason to restrict it only to scientists. All data for DFN-services (availability etc.) are listed in the appendix.

Dialogue Service:
A scientist, who uses the DFN-dialogue service can access to remote computers via specific modules. The mode of the terminals is restricted to line-mode (and not to screen mode!). The module on the remote host side is the input module into time sharing systems. The module on the terminal side (the "packet assembly/disassembly" PAD) can be realized either by hardware or by software which is embedded into the (local) user's host. The data communication is provided by a specific network service that follows a special CCITT-recommendation (X.25). The general situation is outlined in figure 1.

```
        User side                              Remote hosts

    I       +--------+     +---------+     +------+------+
 /-I        I HW-    I     I         I     I      I      I
   I------I PAD     I----I  X.25-  I-----I      I      I
---I        +--------+     I NetworkI     +------+------+
                           I         I
    I       +---+---+      I         I     +------+------+
 /-I        ISW-I  I       I         I     I      I      I
   I-----IPADI    I----I            I-----I      I      I
---I        +---+---+     +---------+     +------+------+
```

Figure 1 DFN-Dialogue

Remote Job Entry:

A user who wants to submit a job to a specific remote machine does
it with the remote-job-entry service of DFN. The RJE-module in the
users computer will send this job to the remote host. The job will
be computed and the output will be routed either back to the user
directly   or to a specific output device (i. e. plotter) which has
to be specified on the job control card. The situation is   sketched
in figure 2.

```
        User Side                         Remote Hosts
        (Host)
                        +--------+      +--------+----------+
                   I        I       I DFN-   I   Input-I
                   I        I-------I RJE    I   queue I
                   I X.25- I        +--------+----------+
        +--------+  INetworkI
        I DFN-  I   I        I
        I RJE   I-------I        I        +--------+  +-------+
        I       I   I        I-------I DFN-   I--IOutputI
        I       I   I        I        I RJE   I  IDeviceI
        +--------+   +--------+        +--------+  +-------+
```

Figure 2 DFN-RJE

File Transfer:

Via DFN-File Transfer files can be sent from a file system in one
computer to a file system in another computer, which normally is
operated under a different system. The files can be sent (PUT-func-
tion) or received (GET-function); two data formats are supported:
binary and character files, figure 3 summarizes the functionality.

```
        Users host                                    Remote host
    +-----+---------+     +----------+       +---------+-----+
    I     I DFN-    I---I X.25     I-------I         I     I     I
    I     I FT      I    I network I       I DFN-FT I     I     I
    I     I         I    I          I       I         I     I     I
    +-----+---------+    I          I       +---------+-----+
          I             .I          I             I
    +---------+     +----------+       +---------+
    I local  I     I          I       I local  I
    I file   I     I          I       I file   I
    I system I     I          I       I system I
    +---------+     +----------+       +---------+
```

<u>Figure 3</u> DFN-File Transfer

Message Handling Service:

The Message Handling service is dedicated to the transfer of data which cannot be handled by conventional file systems, job systems or timesharing systems. Those data (personal messages, messages to groups etc.) need a mechanism that is different from that of the three services sketched above. The user interface is embedded into a functional module named user agent (UA) which specifically does all the local handling (archiving etc.). This module communicates with the message transfer agent (MTA) which is dedicated for example to routing purposes. On the remote side another MTA receives and distributes messages to the local UAs (see fig. 4). UA and MTA on one side are often (not necessarily) embedded into one computer. The overall number of UA's and MTA's are organized in management domains, which normally cover one organisation (legal entity).

```
    +-------------------+     +----------+     +-------------------+
    I +----+            I     I          I     I          +-----+ I
    I I UA I            I     I          I     I          I UA I I
    I +----+--+-----+   I     I          I     I +-----+--+-----+ I
    I         I MTA I I--I     X.25     I--I I MTA I            I
    I +----+--+-----+   I     I network I     I +-----+--+-----+ I
    I I UA I            I     I          I     I          I UA I I
    +-+----+----------+     +----------+     +-----------+-----+-+
```

<u>Figure 4</u> Message-Handling

Graphical Dialogue:

If a user wants to access to data in a remote Graphical Kernel System (GKS) and operate the application on top of GKS from his workstation, this cannot be done by means of the line oriented dialogue, because during the process of communication this sort of service implementation cannot distinguish between graphical output and character output. This is the reason why a specific type of dialogue dedicated to the remote operation of applications using a standardized GKS has been developed within DFN (see fig. 5).

```
   Graph. Workstation                    Remote Computer

+---------------+     +----------+   +------+--------+-------+
I +-----------I     I          I   I      I        I       I
I I GKS-       I------I  X.25    I----I GKS- I   GKS  I Graph.I
I I Dialogue I     I network I   I Dial.I        I Appl. I
I +-----------I     I          I   I      I        I       I
+---------------+     +----------+   +------+--------+-------+
```

Figure 5 Graphical Dialogue

2.2 DFN-Protocol Architecture:

Communication services are provided by means of distributed processes which have to operate in a well defined fashion. The specific rules which define this operation are called protocols. The
ISO (International Standardization Organisation) has defined a model for data communication which layers the modules that are necessary for any communication into seven different layers. The X.25-
network sketched in the section above defines the first three layers. Because the ISO-model is a model for open system interconnection (OSI) it must be possible for each end-system connected to
that X.25-network to be accessed from any other end-system connected to that network. In an X.25-network environment which is a public service in almost every country today, the X.25-access point
is defined by a host specific X.25 address. Another protocol entity on the fourth layer of the ISO-OSI model (the transport layer)
now provides a specific quality for the transport of data. Within
DFN the ISO-Class 0 is taken for the transport layer. On top of
this service the applications (RJE, File-Transfer, Message Handling) reside with their specific needs in data representation. The
overall architecture with the OSI specific layering is summarized
in figure 6.

```
Dialogue    File Tr.      RJE        Mess. Handl.    Graph.    ISO/layer
+--------+-----------+--------+---------------+--------+
I X.3/    I  DFN-     I   DFN- I X.400          I DFN    I
I X.28/   I  FT       I   RJE  I                I Graph. I 5 - 7
I X.29    +-----------+--------+---------------I Dial.  I
I         I  I S O - C l a s s / 0            I        I      4
+--------+-----------+--------+---------------+--------+
I                  X.25                                I 1 - 3
+------------------------------------------------------+
```

Figure 6 DFN-protocol architecture

3. Engineering Principles within DFN-Implementations

3.1 Standardization

The overall principle in all DFN-developments is the use of standardized protocols whenever this is possible. The motivation for
that strategic decision is very simple: Only standardized communication protocols provide communication services between hosts from
different manufacturers in a manufacturer independent way. This is
a direct benefit to the user (whether a computing centre or an end

user) because one will not be restricted to a manufacturer specific communication system. The second benefit for users in using international standards is that the interfaces within the communication systems are well defined on an international level. This makes international communication - which is especially necessary in
the scientific community - much more easy.

3.2 Software Quality of DFN-Products

Users of DFN-products are normally computing centres or end-users
in the scientific community. The requirements for software stability in these environments are very high. The use of operating system interfaces for communication software require professional design and implementation; otherwise no computing centre would use
that software. This problem made it necessary that a well defined
strategy for implementations (in the sense of software production)
had to be adapted. At the beginning of DFN-developments it has
been identified that industrial development procedures should be
used for DFN-products. With the highest priority DFN-projects are
done by manufacturers or software houses. Of course those projects
are carried out on the basis of well defined and documented user-
and architectural requirements.

3.3 Role of Manufacturers

For the same stability arguments as above and for the large problem of software maintenance it is absolutely necessary to involve
the manufacturers in the developement of an OSI-network. Basic stability of communication software can only be received if the product is involved in a very well defined cyclus of software production, maintenance and delivery to the user. Normally manufacturers
have well established means for that purpose. Of course, it is
more easy for most manufacturers to build standardized products -
therefore, the first design principle for DFN (necessity of standardized communication protocols) is the most important one.

4. Application Scenarios

4.1 Remote Job Entry

In Germany an increasing number of vector processors are used in
the scientific community, obviously not every university is able
to operate such an expensive machine. Therefore different groups
of scientists are using a remote vector processor. One example is
the vector processor CD 205 in the university of Bochum which is
accessed from users of five other universities (Düsseldorf,
Aachen, Köln and Bielefeld) via the DFN-RJE service.

The number of calls/month is ca. 3000, the transferred volume of
data is ca. 100 MByte/month.

4.2 File Transfer

A special user group within DFN is a group of electronic circuit design engineers, who is shipping its application data (e. g. circuit layout) with DFN-File Transfer for example from a design site to a host with a specific test software. The number of transferred data is 100 Mbyte/month.

4.3 Dialogue

A lot of scientists are using the line-oriented dialogue in order to access to data bases. Most German information data bases for specific application areas (physics, chemistry, medicine etc) can be accessed via this DFN-service.

4.4 Message Handling Service (MHS)

The implementation of the DFN-MHS which is available today, is installed and used on ca. 20 hosts. This number will increase by a factor of 2-3 when other implementations (see appendix) will be available. Those systems enable the users to send and receive messages not only from German colleagues but from foreign colleagues as well. A gateway to EARN, EUNET and the US-CSNET is provided as a special service.

5. Future Plans

The actual use of DFN-services shows that there is a real demand for such type of infrastructure. A lot of problems (technical and non-technical) have to be solved in the next years. The most important non-technical problem to be solved in the DFN context in the near future is related to financing issues: The service has to be founded by users not by any central agency.

If a review is done of the user scenarios of DFN (see chapter 4) it is necessary to react on growing use of the services. In so far it is natural to deal with the technical issue of high speed data communication (2 Mbit/sec).

Except for those topics which are more speed oriented, DFN has to deal with some architectural developments, too. The first item is defined by the integration of the international standardisation scenario in the data communication field. From an application point of view a standard for filetransfer will be available in the near future - so the DFN-protocol architecture will change in this area.

The second item is related to the development of new distributed applications. If for example several distributed application programs (i. e. data acquisition and evaluation algorithms of a multi-center-study) want to communicate and exchange data cannot easily be handled with the services and the protocol architecture sketched in chapter 2.

It is necessary to develop new application-oriented services and protocols. This will be done within DFN based on two principles: user-orientation (involvement of users) and application of standards (whenever possible).

143

Appendix:

Availibility of DFN-products (October 1986)

	CDC NOS/BE	CDC NOS	CDC NOS/VE	IBM MVS	IBM VM	Siemens BS2000	Siemens MSP	Siemens R 30	DEC RSX 11	DEC VMS	UNIX V	UNIX 4.2	Sperry	ND 100	Prime
MHS	...	...	...	...	IV/86	IV/86	...	...	...	IV/86	II/87	II/87	...	...	...
RJE	A	A	II/87	I/87	IV/86	A	I/87	...	I/86	A	...	...	A	A	A
File Tr.A		...	II/87	IV/86	IV/86	A	IV/86	A	A	A	A	A	A	...	A
T.70	A	A	II/87	IV/86	IV/86	A	IV/86	A	A	A	A	A	A	A	A
PAD	A	A	II/87	A	IV/86	A	A	A	A	A	A	A	A	A	A
X.29	A	A	II/87	A	...	A	A	A	A	A	A	A	A	A	A
X.25	A	A	II/87	A	A	A	A	A	A	A	A	A	A	A	A

Legend:

A: available
II/87: available in the second quarter of 1987

Interactive Data Analysis on Large Data Bases

Techniques to Gain Computational Speed and Ease-of-use

Dr. Richard Reuter & Dr. Rainer Janßen

IBM Scientific Center Heidelberg
Tiergartenstraße 15

Abstract: About 30,000 (June 1986) case histories of kidney transplants are stored in the CTS data base. One purpose of the TRAINS project is to support the analysis of these data. Due to the typical defects of this kind of medical data only "explorative" data analysis seems to be appropriate, which should be done by the medical researcher himself, because he knows about origin and content of the data. In order to make this analysis task as easy as possible from a technical (i.e. data processing) point of view for the physician, he should be provided with an interactive tool. But how should the realisation of such a tool look like ? What are the requirements in detail ? What are the technical implications ? Especially, how can the speed be achieved allowing interactive analysis of a rather large data base. To study this kind of questions a program was designed and implemented for the computation of survival curves and some related test statistics. The chosen statistical methods are only examples to gain experiences for a more sophisticated tool. Some of the experience will be reported and discussed.

1.Introduction

As a part of the TRAINS Project (see the paper of Keppel in this volume) an experimental software tool had to be developed supporting the statistical analysis of the transplant data. A first proposal was the Interactive Statistical Analysis Package (ISAP), which was presented and demonstrated at the International Symposium on Relevant Immunological Factors in Kidney Transplantation, July 1985 in Heidelberg. The underlying considerations and experiences which lead to the development of ISAP are described in [1]

ISAP was designed to enable 'explorative' data analysis by an easy and flexible switch between different methods in order to look at the data from different viewpoints. Although this program was easy to handle by the user (medical researcher) it did not fulfil an important requirement within the whole information system satisfactorily: All the centers reporting their kidney transplantations to the Collaborative Transplant Study (CTS) data base periodically get a report in which among other things hypotheses are discussed which factors seem to be of great importance to the success rate of the transplantation. Essential means in these discussions are Kaplan/Meier Survival Curves which seem to be well understood by the medical research community in the transplantation area. But to detect important factors or to show the effect of some factors to the outcome lots of survival curves are generated and compared.

To generate this or even hundreds of survival curves often grouped around a specific research topic, checked in regular time intervals, ISAP was not appropriate. The intention within ISAP merely was

- pose one question (hypothesis) at a time, explore it with several methods.

Whereas the new requirement is

- pose several/many questions at a time, report the answers given by one method.

In the next sections the design of the new program system is presented with some remarks concerning for possible extensions. The key point is that on the one hand we have to offer an easy-to-use interface allowing to specified the desired 'survival curve' by simple logical descriptions combined with means to administrate large series of such descriptions and on the other hand still wanted to stay fast enough for interactive use. So one of the main issues will be to show how large data sets can be accessed quickly for statistical analysis.

2. The example of Kaplan/Meier Curves

Although the system originally was designed for the production of Kaplan/Meier Curves, we believe that it is an example to demonstrate, how statistical methods could be linked to rather big sets of data and nevertheless could be used interactively. In most clinical trials data analysis is based on at most some hundred cases. Furthermore, only very few evaluations of some statistical tests are required. Under these circumstances computer performance and memory is not a problem at all.

The CTS data base presently contains more than 30000 case histories, and about 10000 new ones are expected per year, each with information on hundreds of variables. In order to enable interactive use of data analysis tools on this large amount of data one really has to care for computational speed and access time to the central data base. So a special concept had to be developed, providing the necessary speed-up and nevertheless general enough to allow the embedding of other statistical methods. The main idea was to do as many precomputations as possible, not to extract all the data needed for a computation at the time when the user initiates that computation, but to extract those data and make computations on those data in advance, whenever one knows that they will be needed.

To give an example: The computation of a Kaplan/Meier curve for a special subpopulation is based on two vectors (see Kalbfleisch/Prentice [2]):

n_j, the number of patients under observation at the time t_j
d_j, the number of failures at the time t_j

These vectors will change for every subpopulation, so that they can't be precomputed. But one can compute two other vectors from which the n_j and d_j can be derived quickly:

R_i, $i = 1,...,N$
T_i, $i = 1,...,N$

where N is the number of cases in the data base.

- If $R_i = 0$, no failure occurred and T_i is the number of days for the i-th patient, telling how long he was under observation.

- If $R_i = 1$, the graft failed and T_i tells how long the graft of patient i functioned.

Now when a subpopulation is specified, the vectors T and R are compressed with respect to that subpopulation giving T^x and R^x. T^x is split in two parts, T_n^x and T_d^x, under the control of $R^x(R_i^x = 0 \text{ or } 1)$. The computation of the frequencies of the entries of T_n^x and T_d^x gives the n_j and d_j. The advantage of the latter is based on the fact, that the direct computation of the vectors n_j and d_j requires about 15 accesses to the data base, everytime extracting as many entries as the subpopulation tells, and that the computations are complicated. If the time consuming extractions and computations are done in advance for all cases, the rest of the computations for a specific subpopulation are cheap (both, in time and space) as described above. More speed-up can be achieved by other precomputations to be explained in subsequent sections.

3. System design

The intention of the system is to support data analysis, hence different subpopulations have to be comparable, i.e. the results have to be computed from identical contents of the data base.

Originally the transplant data are stored in a SQL data base which is updated permanently. Hence the content of the data base may change from one second to the next, which is not appropriate for data analysis.

A first decision was made: In order to guarantee an identical status of the data base for a certain time we do not use the original one. Instead, the SQL data base is converted into a APLDI data base at a certain time. By this all the changes in the SQL data base are ignored until the next conversion. This means that the APLDI data base is fixed for some period. Due to the conversion SQL $\rightarrow$ APLDI, we gain additional advantages:

- The SQL data base contains all cases and variables. While generating the APLDI data base not all cases and variables need to be converted. All those cases may be excluded which cannot be used for the computation of survival curves; e.g. cases for which the reported failure data of a transplanted kidney occurs prior to the transplantation data. A lot of variables in the data base are only of interest for some specialists. Normally they are not required. Besides this, there are some variables, which are sensitive, and which should not be of interest for the data analyst, e.g. names of the patients. So the APLDI data base contains only those entries, which are correct (with respect to the later on computations), and of interest (with respect to the application). Hence APLDI data base is much smaller than the SQL data base.

- The SQL data base organization compromises between conflicting goals like fast access, fast update, small memory requirement. In our application only fast access to all entries of some variables is needed. The APLDI data base is a 'flat' file. The entries of one variable are stored in contiguous records. Thus the access time to all the entries of one variable in APLDI is low compared to the access time in SQL.

- The entries of the SQL data base are retrieved numerically or as characters depending on their definition. In APLDI all entries are encoded numerically which speeds up the later on computations.

When the conversion SQL → APLDI is done, the precomputation of the vectors R and T is performed. But besides this, additional data and variables are prepared to speed up the application. To understand and motivate these precomputations, let us describe a typical application.

As results, graphical representations together with numerical values and certain descriptions are generated. Figure 1 shows the result of a single survival curve comparison. The numerical results are displayed in Figure 2, containing information about the size of the population, confidence interval, the product limit estimator, p-values based on the log-rank test, and some dates giving the data base status. Such graphs may be combined for comparisons of groups. An example is shown in Figure 4. Each of the curves could be described by a subpopulation and a stratification (see Figure 3). So TXNR = 1 ∧ REL = C ∧ MMAB = 0 is the first selection, and so forth.

Actually the user gives for every set of curves he has in mind a description, which is stored permanently (see above). All descriptions are labelled by a name for later references and crossreferences (description, numerical results, pictures). The whole process of generating pictures is split into three steps. First, descriptions of sets of curves have to be given. Second, the user starts the computation by referencing the name of a description and some additional parameters to get the numerical results (including the numerical values of the curves), which are stored permanently.

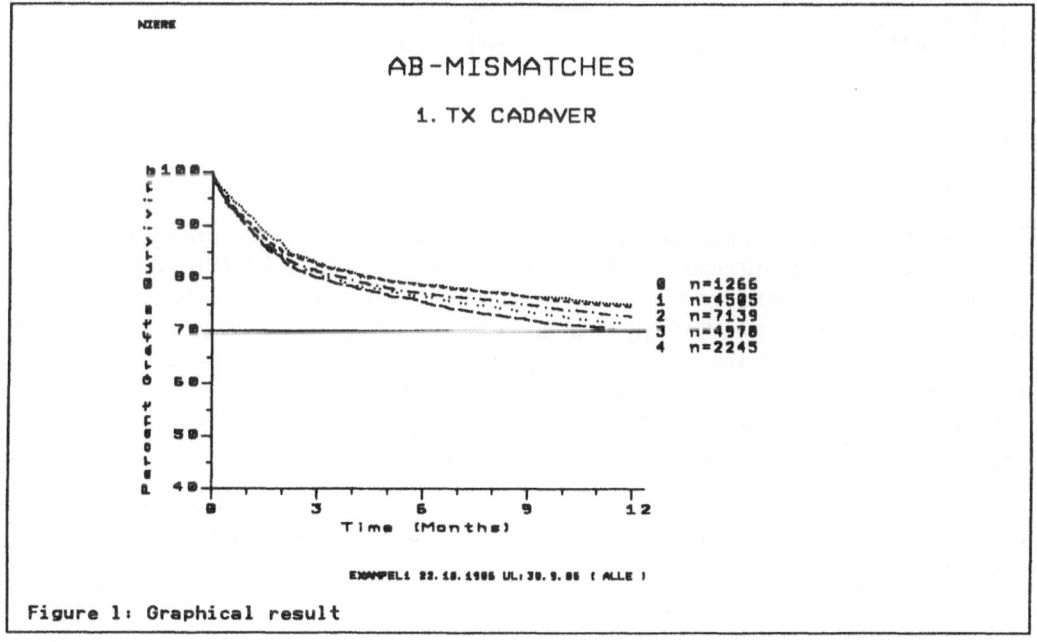

Figure 1: Graphical result

Daten: 30.9.86 (ALLE) Min. = 20 berechnet: 22.10.1986, 17:3:4							

Nr.	1 Jahr	Endwert	SE	95%-KI		Tag	#1	#n
1	75.2	70.5	1.4	67.5	73.2	731	1266	435
2	74.7	68.5	0.8	66.9	70.0	731	4505	1543
3	72.8	66.5	0.6	65.2	67.8	731	7139	2257
4	71.5	64.2	0.8	62.6	65.8	731	4978	1378
5	69.9	62.8	1.2	60.4	65.2	731	2245	611

p-Werte (paarweise) :

	4.1E-1	3.2E-2	1.6E-3	3.3E-4
		3.3E-2	2.4E-4	6.6E-5
			6.0E-2	1.1E-2
				2.9E-1

p-Werte (global) :
2.4E-6

Figure 2: Numerical results

Name:	EXAMPLE1
Kopf:	AB-MISMATCHES
Titel:	1.TX CADAVER
Subpopulation:	TXNR = 1 ∧ REL = C

NR.	Selektion	Legende
1	∧MMAB = 0	0
2	∧MMAB = 1	1
3	∧MMAB = 2	2
4	∧MMAB = 3	3
5	∧MMAB = 4	4

Figure 3: Picture description

Third, the user starts the generation of a single, double or triple picture by referencing 1, 2 or 3 names of numerical results. The pictures are labelled by the names of the involved numerical results and stored permanently. Every step only needs some seconds (the actual 'elapsed' time strongly depends on the workload of the whole machine). Computing and storing the result of a typical description of a set of curves (4 to 5 selections, composed of 10 to 15 simple selections, survival curves for one year, with possibly several thousand patients in each selections as in the examples in Figure 1,2,3) needs about 2 CPU seconds on the IBM 4381 M2.

In order to achieve this speed, a special technique is required. A first version of the program was used for about one year. By inspections of the user-defined descriptions of sets of curves we found

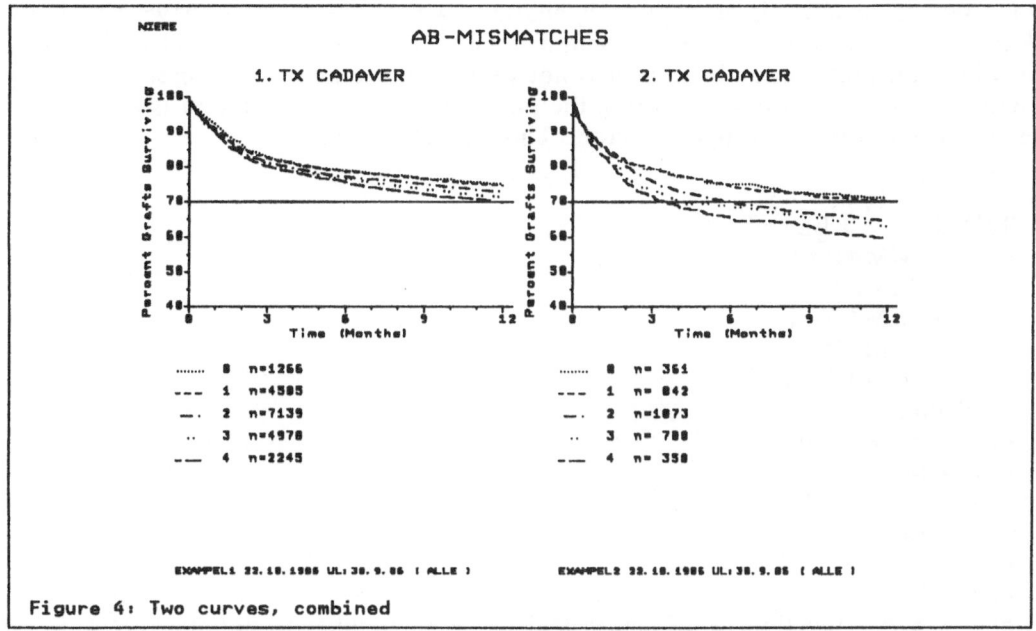

Figure 4: Two curves, combined

- some hundred descriptions

- consisting of some thousand selections (a selection is a subpopulation plus one part of a stratification), 4-5 per description

- each selection is composed of - in average - 10 to 15 'simple selections' like TXNR = 1 or REL = C (a simple selection is a selection without any logical function symbol like ∨ , ∧ , ⩔ , ⩘ , ∼).

- the pool of all selections in all descriptions is composed of only a relatively small number of simple selections: about 400 (and not '500 descriptions times 4 selections times 10 simple selections equals 20000 simple selection').

Every simple selection defines a logical vector, whose i-th entry tells, whether case i belongs to the selection or not. Thus every logical vector defined by a selection (as a logical expression of simple selections) could be computed very quickly, if the logical vectors of the involved simple selections are known.

As mentioned above the vectors R and T are precomputed. So the troublesome part within the evaluation of a numerical result was the extraction of the logical vectors defined by the selections. Although we decided to use a APLDI data base within our system with all its advantages, the direct extraction of the variables (10 to 15 in a typical description) involved in a selection became more and more time consuming when the number of cases was growing. When we started 13206 cases were stored and an early version of the program worked fine. But with about 25000 cases the response times of the program became too long (about 20 CPU seconds for a typical description), so that we could not speak of 'real' interactivity any longer. Because of the remarks above, we decided to evaluate all logical vectors of the simple selections in advance and to store them permanently. So when the user defines a new description

of a set of curves, this description is analyzed with respect to the known simple selections. If an unknown one is detected, its logical vector is evaluated and stored. This
is a little bit time consuming, but it is not very often necessary, if the user has already
defined a lot of descriptions. During the phase when a new description is analyzed, it
is translated in terms of logical vectors. This translation is stored too.
Example:
Description (selections)
  TXNR = 1 ∧ REL = C
        ∧ MMAB = 0
        ∧ MMAB = 1
        ∧ MMAB = 2
        ∧ MMAB = 3
        ∧ MMAB = 4
Translation:
  C ← (S 1) ∧ (S 5)
  C ∧ S 14
  C ∧ S 15
  C ∧ S 16
  C ∧ S 17
  C ∧ S 18
where S x is a function, which reads logical vector number x. For the translation a
table is inspected giving the simple selections and the numbers,e.g.
1 TXNR = 1
5 REL = C
14 MMAB = 0
15 MMAB = 1
and so on...
This scenario is summarized in Figure 5.

The user only knows objects he has defined or initiated himself: descriptions of sets
of curves, numerical results, graphical results. To handle all these objects, the user is
supported by an administrative system, in order to create, change, copy, edit, delete
objects or to use various output devices (line printer, color printer, laser printer, plotter, display). The split into the three objects (description - numerical results - pictures)
enables us to consider further possible improvements in the future. Principally, different parts could be done in parallel or interleaved. For instance, when the computation
of numerical results, i.e. the panels for the subsystem for numerical results are replaced
by the one for the graphical results. Within that time the actual computation could
be done on a second processor. Even if no second processor is available, one can exploit this potential for parallelism on a single processor by an additional virtual machine.

The user program has some additional features.

1. One can create lists, e.g.
    MMABTXNR   Name of the list
      MMABTX1
      MMABTX2   Names of descriptions
      MMABTX3   of sets of curves
      MMABTX1 MMABTX2 MMABTX3

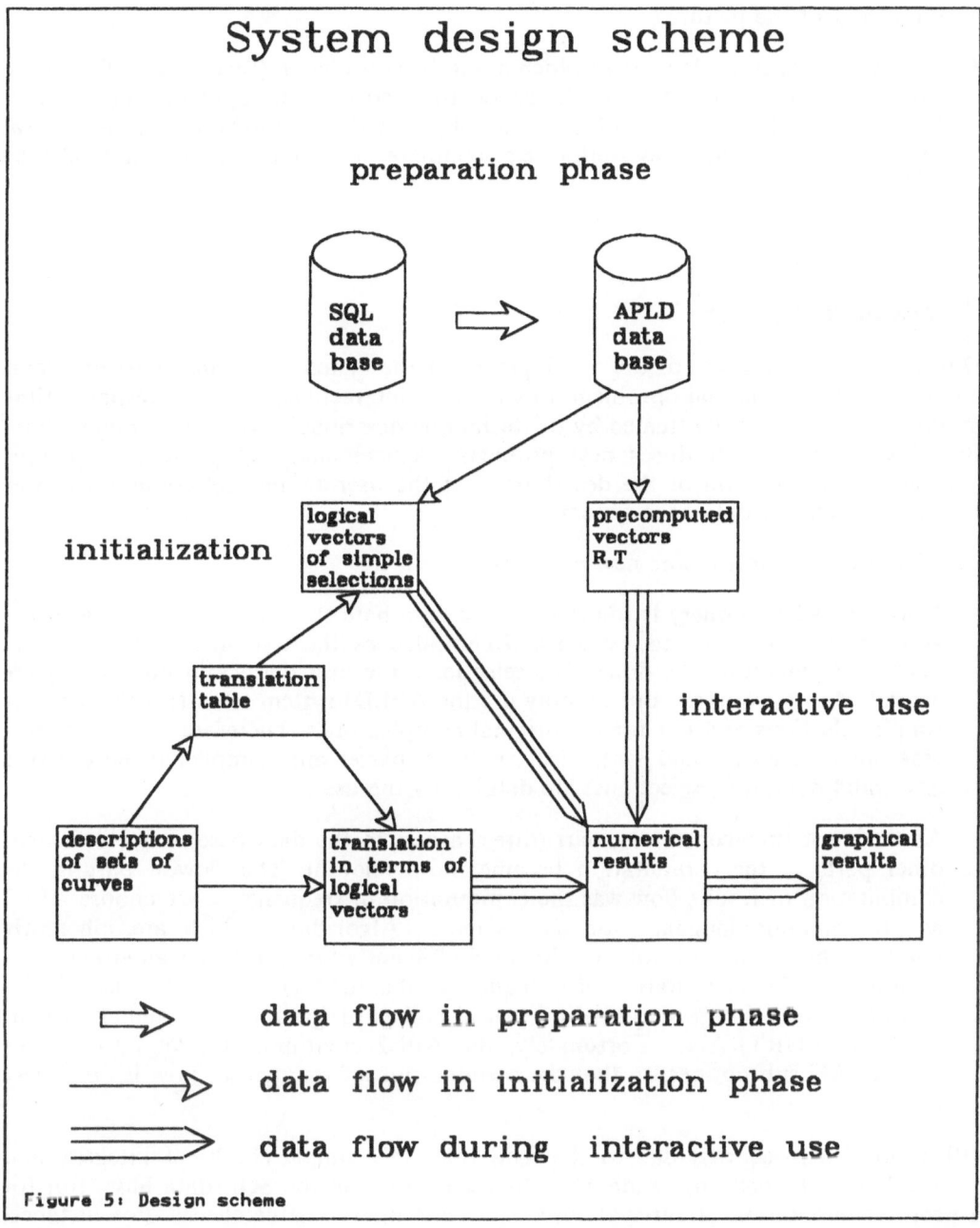

Figure 5: Design scheme

Once a list is 'started', the numerical results of the descriptions, contained in the list, are computed, the graphical results are produced and both stored and optionally printed. (The list, shown above, would produce 3 sets of numerical results and 4 pictures: 3 single, 1 triple picture)

2. At the beginning and during a session the user can set some global values which affect for instance

- the layout of the picture

- the time of the transplant dates which are to be considered. (By changing the global value, it is possible to compare the curves for a special subpopulation in a certain time interval with the curves of the same subpopulation but in another time interval using always the same description; e.g. compare all first transplants in 1982-1984 with those 1985-1986)

- etc.

4. Technical problems

During the whole process of the development of the system some technical problems had to be solved. The main problem, to guarantee interactivity (i.e. keep response time minimal) was successfully treated by the technique described in the preceeding section. But this technique introduces new problems (lexical analysis, syntactical analysis, parsing and compilation of the descriptions of the user-defined selections must now be done outside the data base system)

But this technique introduces new problems.

- Normally, when a query is transferred to a data base system, that query is lexically and syntactically analyzed, parsed and compiled by the system, and then the real data extraction could be done. The selections, the user defines in our system are actually APLDI queries. But we only use the APLDI system to extract the data for simple selections and not for the original complex ones. Therefore we had to provide our own lexical and syntactical analyzer, parser and compiler to ensure that only valid queries (=selections) are defined by the user.

- As the most time-consuming part (direct access to the data base) was eliminated, other parts of the computation became more relevant. The slowest part in the computation of results now was the computation of frequencies. We choose APL2 as programming language for our system. Algorithms, which are inherently parallelizable or vectorizable, are handled efficiently by APL2. But an efficient algorithm, for the computation of a frequency table for a given vector, especially for vectors of short integers, typically is serial and is programmed adequately for instance in FORTRAN. Fortunately, the APL2 environment allows to involve FORTRAN subroutines, so that the computation of frequencies now is done very fast.

Other problems occurred due to the fact, that only simple checks of integrity and plausibility of the data are made when they are stored in the SQL data base. But for a proper computation of survival curves we need inter-variable checks. For instance, if a failure is reported 5 months after the transplantation date, the variable UPD6MO (clinical update grade after 6 months) must contain one of the 'failure characters'. Another problem: what has to be done, if a failure is indicated in UPD6MO but the failure date is incomplete or missing?

A problem of quite another kind is the following: The legend texts of the curves in a single picture should be placed at the right end of each curve. But where should they be located, if the end points of some curves nearly coincide? A proper placement can be found by solving a quadratic optimization problem. As the exact solution of that mathematical problem can be too time-consuming, a semi-optimal algorithm was developed.

5.Conclusions

The developed system allows in an interactive way the analysis of the data by one method. By the special techniques used it can serve as a model for a more sophisticated system from a statistician's point of view. We have shown that a statistical method can be used interactively and effectively and that the handling could be user-friendly, although a huge amount of data has to be treated. This is merely achieved by doing as much precomputations as possible. The flexibility of the design was proved recently, when the program was extended in order to connect not only to the kidney transplant data base but also to a heart transplant data base. Additionally we can now handle graft and patient survival curves. The latter caused some problems, since the structure of the variables, from which the information on survival and observation times is drawn, is totally different to the corresponding variables for graft survival. So despite the fact that the statistical method is the same for graft and patient survival, namely the Kaplan/Meier product limit estimator, the implementation relying on the data structures and on consistency rules is quite different.

References

1. R. Janßen, R. Reuter: Interactive statistical analysis of transplant data; in: H.J. Jeschinsky, H.J. Trampisch (eds.): Medizinische Prognose - und Entscheidungsfindung, Springer 'Medizinische Statistik und Informatik' Vol. 62 (1985)

2. J.D.Kalbfleisch, R.L.Prentice: The statistical analysis of failure time data; Wiley (1980)

MODELLING HUMAN ORGANS
APPLICATION TO KIDNEY ANATOMY

Jean SEQUEIRA
Centre Scientifique d'IBM-France, PARIS

Pierre CONORT
Hôpital de la Pitié-Salpétrière, PARIS

Abstract

New medical imaging techniques, such as X-Ray Scanning or Magnetic Resonance, give us the capability of examining live organs without having to use surgical processes. Most of these techniques produce images of organ sections. Medical scientists can use such a set of images to figure out the three-dimensional shape of organs.

During the last ten years, algorithms have been developed to display such a set of data in three-dimensional space. Realistic views of organs can be computed by using these algorithms. But it will be difficult for the physician to use these data for purposes other than visualization.

In this paper, we propose an original solution to this modelling problem. By using a set of shape primitives and operations on these primitives, we allow the user to interactively create a three-dimensional model suitable for his application.

We are developing a system based on this approach to model kidney cavities. Geometrical parameters of these cavities can be extracted from the models generated by the medical scientist.

Introduction

Computer applications in medical care have been greatly increasing during the last ten years. Combined with other electronic devices, computers can produce images which represent human organ sections. Such a way to get informations on patient organs widely improves diagnosis and surgery efficiency. But we can go through a new step by generating three-dimensional models of these organs and by displaying them. Most of research in this area focuses on the visualization process. But, in order to efficiently exploit the data collected and processed by the computer, we need to create a high-level three-dimensional model of the organ to be displayed. An interactive approach to get such a model is described in this paper as the way to use it for the study of kidney anatomy.

I. 2D and 3D data visualization in medical care

Classical X-ray radiographs give us a projection of human body inner parts, with an enhancement of high-density elements. But they cannot give us a complete view of organs, such as in cross-sections. Recent imaging techniques solve this problem, usually by computing those sections from a set of projections along different directions. Physicians can then get a full examination of organs by using such equipments as X-ray scanners or those producing Magnetic Resonance, ultrasonic or radionuclide images. The information collected on the organ (density, acoustic property, etc.) depends on the type of equipment but, in all cases, it is represented as a set of local information and displayed as organ cross sections.

In order to get a better visualization of organ shape and configuration in space, we need to structure this local information into a three-dimensional model and to display it on a graphics device. Such a display allows the physician to estimate the data three-dimensional coherence. By rotating and zooming this model, changing some of its parameters, as colors, thresholds or clipping boundaries, he can emphasize some of its basic elements.

The classical method to produce such images consists in going through three steps which are, sequentially:

1. Image processing

2. Modelling

3. Display

The image processing step produces a set of relevant two-dimensional features which are either areas (sets of pixels) or contours (sets of edges bounding areas). Depending on the type of data generated at the previous step, we can define a voxel-based model (in the case of areas) or a polygon-based model (in the case of contours). Let us describe these two ways of modelling and displaying three-dimensional medical data.

Pixels defining areas in parallel cross sections are associated to volume elements called *voxels*. We can then provide a spatial structuring of these voxels by using *octrees*. An octree is a data structure in which each node has eight sons. The root represents a cube which contains the whole scene to be displayed. We then subdivide this starting cube into eight identical ones: they represent the first level nodes. We continue this recursive structuring until cubes are full or empty, (all voxels contain the same information). The octree data structure is suitable for a fast display of a scene defined as a set of voxels.

In the case of contours, we tile the model by selecting points on these contours and by joining them with edges. We then get a set of polygons which are triangles in most cases, or else four-sided non-planar polygons. We note that finding corresponding points on two following contours is not an easy task: it is still an unsolved problem in many situations although recent algorithms bring some solutions. Many efficient algorithms have been designed for displaying sets of polygons with hidden-part removal (and shading).

II. Modelling

In both cases, we cannot get well-structured models. In addition, these modelling approaches generate large sets of data, and thus, they involve very large amounts of computation.

Displaying three-dimensional representations of organs is an important improvement in the area of medical imaging. But it is much more important to have a highly-structured model which can be exploited efficiently by the user. Data, as defined before, cannot be related efficiently to medical entities: pointing out a polygon or a set of voxels will not allow the physician to pick and study a relevant medical element as an artery, a kidney cavity or a tumour.

In order to get such a structured three-dimensional model, we do not let the computer create it automatically but we use an interactive process which takes advantage of the physician's knowledge. This modelling approach allows the user to make up consistent models suitable for specific applications, as the study of a particular part of an organ. By using a highly interactive communication with the model, the medical scientist can easily analyse its geometrical parameters, such as curvatures, twistings, diameters, lengths, surfaces, volumes, etc. In addition, this modelling can be used as a basis for the comparison and classification of models.

III. A Computer Aided Design Solution

The basic idea of the solution we propose is to let the user interact with the system so that he can drive the modelling process, especially by taking decisions in critical situations. This control occurs at every step of the process and allows the physician to select geometrical primitives to make up a highly-structured model. These primitives are designed to model volumes bounded by free-form surfaces. Each of them is associated with a particular topological feature.

By using image processing procedures, we can reduce the initial data set to relevant features as contours or areas. The physician can then select those which are significant for his application. These selected data are used as a visual support for the modelling step: looking at them in the three-dimensional space helps the user to choose the right primitives and to adapt them to these data. The system then provides automatic deformations of these primitives until they fit precisely to the selected data.

III. 1. *Data Selection*

Classical three-dimensional modelling and display, as described in the first paragraph, need initial data presented as parallel sections of the organ to be displayed. With the approach described in this paper, that restriction vanishes. This is especially interesting if the data come from an ultrasonic or radiographic equipment.

A first data selection is provided by classical image processing computations. Contours can then be selected by pointing out sets of points which interest the physician. In order to do that, he has to manipulate closed curves until they roughly fit these groups of points. The system then optimizes the shape of the curves. These curves are cubic B-Splines and are manipulated through global operations on their control points.

III. 2. *Primitives to model free-form objects*

Free-form primitives are the basic elements of this modelling process. They enable the user to make up models of volumes bounded by free-form surfaces, as in the case for organs. Primitives are defined by their topological characteristic which can be simple (e.g. to define a single patch or a tube) or very complex (e.g. to define cavity networks). They must be adaptable to any kind of geometrical situations.

A free-form primitive is represented by a set of free-form surfaces (Bezier, B-Splines, Beta-Splines, etc.) and relations between their control points. The type of surfaces defines local properties and continuity characteristics of the primitive. Relations between control points define its topological feature. For example, we can define a tube as a single cubic B-Spline patch in which we duplicate the three first control points columns (or rows) onto the three last ones.

III. 3. *Operations on primitives*

In order to modify their shape, we need to provide operations on these primitives. The user will then be able to manipulate them until they roughly fit the selected data (cubic B-Spline curves defined at the previous step). We could provide deformations on these primitives by moving their control points. But this solution is not acceptable for a friendly interface: it would be a long and boring means to do manipulations such as those required in this modelling problem. In addition, the user is not supposed to know the mathematical definition of primitives. The ideal solution would be to define transformations which let him manipulate primitives as "if he had them in his hands". We have gathered operators into three classes which represent three communication levels:

- The first one consists of basic operations which are applied either on a single control point or on a group of control points.

- At the second level, the user does not need to know the mathematical definition of primitives. He directly manipulates entities related to their geometrical properties (e.g. curvatures, twistings, central axis, etc.).

- Third-level operators have to simulate mechanical actions on primitives in order to provide a very friendly communication between the user and the system.

III. 4. *Primitive automatic deformations*

After the initial data selection, the user has picked free-form primitives in a geometrical knowledge base to define a well-structured three-dimensional model. It is then necessary to optimize the shape of these primitives until they fit precisely to the selected data. Automatic deformations are provided by the system by means of control point translations so that model elements attain the suitable shape.

III. 5. *Display and interactivity*

Getting a visual feedback of all operations is the key point of the approach described in this paper. Thus, we need to represent three-dimensional shapes in a way compatible with interactivity. We use three levels of representation depending on the desired response time and image quality:

- Wireframe display, without hidden-part removal, enables real-time manipulations, as looking at the model from many viewpoints or deforming primitives.

- A specific algorithm has been designed to get fast realistic displays of the model. Although this algorithm produces low-quality shaded images, it can widely reduce ambiguities in model understanding.

- Classical hidden-part algorithms, as ray tracing for example, can produce highly realistic images.

Viewing parameters are interactively modified by using a tablet with a cursor:

- The cursor location on the tablet is related to a view point position on a sphere: motion in X corresponds to displacements along parallels, and motion in Y relates to displacements along meridians.

- Pushing cursor buttons can be used to increase or decrease the sphere radius, or change a scale factor.

III. 6. *Conclusion*

This Computer Aided Design solution gives the physician the capability of getting a well-structured, compact and accurate model of an organ. The model structure enables a complete study of its geometrical parameters. In addition, this modelling process allows the comparison between geometrical characteristics of such models, stored in a data base (e.g. for classification).

IV. Application to Kidneys

Current techniques in renal surgery and endoscopic diagnosis and treatment (calculi, tumour, etc.) require a good knowledge of renal pelvis and calyces orientations, sizes and diameters. It is especially the case of kidneys which have been grafted in iliac fossa. Although renal calculi or excretory tumour are rare in kidney transplants, no failures, such as infection, haemorrhage or residual calculi, are acceptable.

It is then necessary to follow up the evolution of grafted kidneys. Nowadays, urography (IVP) is the only imaging technique which provides enough precision to do it. But it only gives projections of kidney cavities and we need a three-dimensional interpretation of these urographies to analyse kidney geometrical parameters. Such interpretations can be obtained by using the approach decribed previously.

Three-dimensional modelling and display of kidney cavities will allow a simulation of the endoscopic exploration. In addition, it will help urologists in teaching and in training before actual operations. There are many other possible applications in urology such as help in renal surgery indications and kidney puncture, improvements in fiber-optic endoscope design and kidney cavity analysis.

V. Bibliography

1   Bagley D.H., Rittenberg M., Huffman J.L., Lyon E.S.
     Flexible Ureteropyeloscopy: Reaching the Inferior Infundibula
     20th Congress of the International Society of Urology, Vienna, 1985.

2   Barnhill R.E., Boehm W.
     Surfaces in Computer Aided Geometric Design
     North-Holland, 1983.

3   Barr A.H.
     Global and Local Deformations of Solid Primitives
     Computer Graphics ACM, 1984.

4   Chiyokura H., Kimura F.
     Design of Solids with Free Form Surfaces
     Computer Graphics ACM, 1983.

5   *Computer Graphics in Medicine and Biology*
     IEEE, Computer Graphics and applications,
     August 1983.

6   Conort P., Jardin A., Hureau J.
     Pyeloscopie Expérimentale. A propos de 38 cas.
     Communication à la Société Anatomique de Paris, 23 Novembre 1984.

7   Foley J.D., Van Dam A.
     Fundamentals of Interactive Computer Graphics
     Addison-Wesley, 1982.

8   Fuchs H., Kedem Z.M., Uselton S.P.
     Optimal Surface Reconstruction form Planar Contours
     Comm. of the ACM vol.20 n.10, Oct. 1977.

9   Ganapathy S., Dennehy T.G.
     A New General Triangulation Method
     for Planar Contours
     Computer Graphics ACM, 1982.

10   Keppel E.
     Approximating Complex Surfaces
     by Triangulation of Contour Lines
     IBM J. Res. Develop., n.19, January 1975.

11   Kochanek D.H.U., Bartels R.H.
     Interpolating Splines
     with Local Tension, Continuity and Bias Control
     Computer Graphics ACM, 1984.

12   Meagher D.
     Geometric Modeling Using Octree Encoding
     Computer Graphics and Image Processing,
     Vol. 19, June 1982.

13   Meagher D.
     Octree Generation, Analysis and Manipulation
     Technical Report IPL-TR-027,
     Image Processing Laboratory,
     Rensselear Polytechnic Institute,
     Troy, N.Y., September 1981.

14   Sequeira J.
A Computer Aided Solution
to a 3D Modelling Problem
International Conference
on CAD/CAM/CAE for Industrial Progress,
Bangalore (Inde), 29-30 Juin 1985.

15   Yamaguchi K., Kunii T.L., Fujimura K., Toriya H.
Octree-Related Data Structures and Algorithms
IEEE Computer Graphics and Applications,
January 1984.

16   *Compte rendu de la journée*
"Images Médicales et Visualisation 3D"
Centre Scientifique d'IBM-France,
13 Novembre 1985.

Donor/Recipient Matching in Kidney Transplantation

Dr. Rainer Janßen & Dr. Richard Reuter

IBM Scientific Center Heidelberg
Tiergartenstraße 15

Abstract: Since kidneys available for transplantation are a scarce good it seems clear that they should be put to 'optimal' use. But it is far from clear what 'optimal' means. Even if we restrict ourselves to 'optimal success rates' - and it may be doubted that this is the only relevant matching criterion - several definitions are debatable.

This paper describes some, basically mathematical, tools which can contribute to a more precise understanding of this notion of optimality. It also points out limitations of present data collections - as far as known to the authors - in providing the information needed to develop a sound matching procedure.

In a final section some consequences are drawn for the design of a computer aided decision support system for this matching problem. The main concern here is not the user interface, be it an expert system like tool or a simple menu or whatever, but the requirements for a computer network and a distributed data base for this system.

1. Introduction

Transplantation of kidneys has become an accepted and widely used treatment of patients with end stage renal failure. Since kidneys available for transplantation are a scarce good compared to the large number of patients waiting for a suitable kidney to be implanted it is important to put them to optimal use. This is easily stated, but hard to do. Despite immense research efforts over the last decades in clinical and theoretical transplant immunology there are still many controversies even w. r. t. some of the most important factors influencing the outcome of a transplantation. And the situation is even worse, up to now there is no precise definition what is meant by 'optimal'. Of course, it is commonly accepted that the probability of a successful outcome of a transplantation should be an important part of the donor/recipient matching procedure. Again, there are different meanings of this seemingly simple notion of successful outcome and we will discuss some of them in detail.

Taking the estimate of success rates as the basic decision criterion leads to pitfalls well-known to so-called greedy optimization techniques. And there are more side effects and limitations to be considered. Furthermore, restrictions are imposed by present data collections (as far as known to the authors). They will be discussed in the fourth section, and ways to improve upon the situation will be shown.

A strategy will then be discussed to develop a matching procedure circumventing previously described pitfalls. The mathematical toolskit will be sketched. Since an actual implementation of such a procedure has to take into account the dispersion of the people and the information involved in the decision process over many transplant centers and also different departments in the respective centers some system aspects will be discussed in the last section. Mathematically interested readers are referred to the appendix for details on some of the techniques involved, especially on open research problems.

2. Different notions of successful outcome and optimality.

Of course, a basic ingredient of any definition of a successful transplantation is that the recipient is still alive. And in extreme situations this might be the only interesting question. But it is not only the fact of staying alive which counts, it is also the quality of life, especially the function of the transplanted kidney. So instead of patient survival we might look for graft survival. In order to be able to state the different criteria more precisely let us fix some notation: We assume henceforth that the donor and recipient are described by a fixed set of variables

$$D = (D_1, \dots, D_n), R = (R_1, \dots, R_m)$$

Specific values of these variables are denoted by small letters

$$d = (d_1, \dots, d_n), r = (r_1, \dots, r_m)$$

For most of our purposes D, R are viewed as random selections from the population of all donor/recipient combinations (whatever that means, the mathematically inclined reader will easily fill in the details on measure spaces etc.) The patient rsp. graft survival time $T_p(D,R)$ rsp. $T_g(D,R)$ are now positive real-valued functions. Here graft survival means, that the patient is not back on chronic dialysis, as in the Collaborative Transplant Study (CTS). Furthermore let

$$P(T_p(D,R) \geq t): = probability, that \ T_p \ exceeds \ the \ fixed \ time \ t$$

$$E(T_p(D,R)): = expectation \ of \ T_p$$

$$P(T_p(D,R) \geq t/D = d, R = r): = conditional \ probability$$

$$E(T_p(D,R) \geq t/D = d, R = r): = conditional \ expectation$$

(analogously for T_g). It is clear that

$$P(T_p(D,R) \geq t) \geq P(T_g(D,R) \geq t)$$

for all D, R and t, but there is no linear or other regular dependency between these values. It might well be that e.g. a certain drug improves the graft survival and at the same time increases the lethality rate. Another example is the age of the recipient: due to reduced physical fitness the percentage of deaths following an immunological rejection is larger for elder patients even if the graft survival rate were the same as for younger patients. So there's a real difference in trying to increase T_g instead of T_p .

Hence a choice has to be made and this is mainly the job of the physicians. Nevertheless, here are some arguments leading us to focus our emphasis on T_g :

1. The lethality rate $L = 1 - P(T_p \geq t)$ consists of three main factors

   $L = L_0 + L_f + L_t$

   L_0 is the basic operational risk, depending mainly on the physical fitness, L_f is the risk to die in case of failure, L_t is the risk of a specific treatment. L_o is a kind of a base risk, any recipient has to accept, and L_f is related to what is measured by T_g. The assumption is that L_t is probably comparably small to the other summands in most situations.

2. The main interest of the patient is to cure the disease not just staying alive under often very restrictive conditions. (This holds true for society, too; but we do not want to dive into social economics here).

3. Looking for T_p introduces a strong bias against elder recipients. If this is really felt necessary, for whatever reason, it should be stated elsewhere explicitly.

For this reason let us stick for the following discussion to T_g .

Due to random fluctuations in the survival time, even if we fix the parameter values $D = d$, $R = r$ rsp., it is not meaningful to try estimating T_g . Instead, we look for some probabilistic expression based on T_g. The most natural choice is $E(T_g(D,R)|D = d, R = r)$, something like the average over the survival times for specific parameter choice d, r. But this choice is rather impractical because it is almost impossible to estimate this value based on existing data collections. The reason is that it is strongly influenced by the 'long survivors' and since transplantation is a relatively new technique there are not so many old enough data collections to estimate this distribution correctly. Even if we continue to collect data for another ten years it is doubtful whether they'd be useful anyway. Since medical knowledge and experience is growing rapidly, conclusions drawn from data older than five years are to a large part irrelevant or misleading for present circumstances. There need not be drastic changes like the finding of the HLA - DR or a new drug like Cyclosporine; just the combined effect of seemingly small changes in the experience of surgeons and clinical staff, improvements in the quality of HLA-typing, better knowledge on the proper dosage of new drugs, new techniques and tools for the post - operative surveillance to detect beginning rejections timely and precisely, and so on. Hence estimators for success rates based on more than five year old data are not reliable. Due to some delay in the reporting procedures and since we need the data of a large enough time interval to get large case numbers and hence small confidence intervals it is in general not meaningful to follow graft survival curves for more than three years.

Our proposal is to base the development of a matching procedure on the conditional probability

$P(T_g(D,R) \geq 1\ year|\ D = d, R = r)$.

for the following reasons:

1. It takes into account the short - term rejection.

2. Only very few examples are known where a crossover of survival curves after one year has been observed.

3. It is easily and timely adapted to changing circumstances.

4. It can be defined and measured quite precisely.

A first optimization procedure then is:

Given a donor D with specific parameters d, find a recipient R on the waiting list WL with parameters $r_{opt.}$ such that

(P1) $\qquad P(T_g(D,R) \geq 1 | D = d, R = r) = \max_{R \in WL} P(T_g(D,R) | Dd, R = r).$

It should be pointed out that similar to the earlier mentioned effect   of replacing T_p by T_g this notion of optimality compared to $E(T_g(D,R)|D = d, R = r)$ reduces the bias against elder recipients. The short time horizon eliminates the effect of the generally, ill or not, shorter rest life time of elder people. Other biases and side effects introduced by this optimality criterion will be discussed in the next section. (For further remarks on the conditional probabilities in (P1) see Appendix B).

3. Side effects and limitations

The first thing happening to any general rule is that somebody comes up with 'exceptions'. Typical examples here are the urgency list (recipients who need a graft within the next weeks or they will die) or the HIT list (recipients who are sensitized against a high percentage of randomly selected donor tissue). Patients on these lists will receive an available donor, whatever their estimated survival time is, if only they have a negative crosssmatch test against the donor's tissue. In the long range, with increasing numbers, these special lists may have a negative effect on the average success rate. This will be demonstrated by two examples:

Example1: Blood group 0 and rare antigens.
Since recipients with blood group 0 or rare antigen combinations have smaller chances to get a good match, (P1) has a strong tendency against these patients. They will stay on the waiting list too long, finally drop out to one of the special cases lists, and here they get a transplant under considerably worse conditions.

Example 2: Selection hypotheses for the transfusion effect.
The selection hypotheses (see e.g. Mickey [13]) tries to explain the beneficial effect of pretransplant blood transfusion at least partially by the fact that mainly the so called high responders, reacting strongly against foreign tissue, will be sensitized, so they won't pass the crossmatch test and won't be transplanted. So the success rate will be increased, but in the long range the sentized patients will be put onto the HIT list - and there eventually receive a kidney under worse conditions than without being sensitized. (The selection hypotheses provides also a typical example of a treatment where the opinions of patient and physician might differ, see Appendix F for a discussion on the decision tree).

Both examples show the disadvantage of a too greedy optimization. This 'greedy optimization effect' is illustrated by the following example:

Example 3: Greedy optimization

Let us assume there are two recipients waiting and two donor kidneys arrive. The respective success rates (in percent) can be read from the following matrix:

	D_1	D_2
R_1	90	80
R_2	70	40

With the greedy procedure (P1) we would match (R_1,D_1) and (R_2,D_2) giving an average success rate of 65 instead (R_1, D_2) and (R_2, D_1), giving an average of 75. So it is necessary to take into consideration possible long term effects. A mathematically interesting formulation has been proposed by David, Yedical, (see Appendix E). But there we run again into the same problems as before: it is hard to collect long term follow- up information, and at the point in time one gets them they are probably meaningless anyway. A heuristic procedure to balance the bias against groups with worse chances for a good match and to dampen the effect of the short-sighted greedy procedure (P1) will be discussed later.

Another problem, we will touch upon here only lightly, is which variables d_i, r_j to include in our estimation procedure (P1). It is easy to say that those variables should be included which result in the best - fitting estimator for the success rates but it must be kept in mind that donor/recipient matching bears also kind of political aspects: society has to agree with the decision procedure, and even more important, the transplant centers and the physicians supposed to - at least in principle - obey the rules of the general procedure have to agree. It is e. g. known that the race has an impact on the outcome but we doubt very much that it is acceptable to include that variable in a general matching procedure.

Much more controversial is probably the opinion w. r. t. the so-called center effect which is supposed to measure to some extent the quality of the transplant center. First, there is up-to-now no precise definition of this effect and how to measure it. Most existing analyses look like self fulfilling prophecies, the center effect is there defined over the average success rate of the respective center and hence has a good prognostic value, especially if tested on the same set of patients on which it was 'measured'. Furthermore, the center effect is not an independent variable. It depends very strongly e. g. on the selection of the recipients: a center not transplanting elder or diabetes mellitus recipients in general has better success rates. And if it is desirable to enforce a specific selection process this should be formulated explicitly by using the independent variables disease and recipient's age instead of implicitly over the center effect. Second, the question arises if it is practically feasible to include this effect: wouldn't it just result in excluding some centers from the general matching procedure?

As a last example let us consider a variable from Collaborative Transplant Study called evaluation. Here the physician in charge is asked to give an estimate of the recipients chances (good, medium, poor). Since it correlates nicely with the success rate it would potentially improve the quality of our estimator but due to the present data collection procedure - the forms are filled out after transplantation - it is unclear how much post - transplant information is contained in this estimate. Hence it is doubtful if it has the same prognostic value in a matching procedure where it has to be defined weeks in advance. This leads us to problems with present data collection procedures which will be discussed in the next section.

4. Some consequences for the data collection

There are many transplant registries and probably as many data collection procedures around. The intention of this section is not to describe design criteria for the 'general' registry (whatever that is) but for a registry supporting the development and continuous checking of a matching procedure. The two basic questions are:

- Which data do we collect ? Meaning: which variables, how are they measured, during which time ?

- How do we collect the data? Meaning: how do we organize, administrate and computerize the data collection?.

Figure 1: Data quality

The different views of the data in multicenter studies are presented in figure 1 (top-down):

- such a large set of beautiful data, it will help us resolve all open question on relevant factors for transplantation

- but a closer look shows that there are holes: a lot of missing entries even sex, age or blood group of patients are sometimes missing

- and then some data look rather weird due to typing errors, missing plausibility and inconsistency checks

- finally, several variables are not precisely defined, even the clinical outcome grade, the most important follow-up information, so we feel the data rest on shaky grounds.

What can be done to improve this picture? Part of the problems can be dealt with by online data entry directly at the transplant centers and the use of computer networks. These system aspects will be studied in other papers presented in this symposium, so the 'how to collect' question won't be studied here. But which data should we collect?

1. Data containing pure post-transplant information can't be relevant for the matching: so it's probably irrelevant to collect the drug protocols which are not known before transplantation, but the cold ischemia time can at least be roughly estimated before transplantation.

2. Collect only the generally accepted variables, like age, disease, HLA typing, blood group, ischemia time, number of transfusions, and several others. Research might from time to time lead to the exchange of some of these variables. But any participating center must submit the information on these basic variables, else it will have to leave the study.

3. Include only precisely defined variables. In a data base for research purpose we might include variables like evaluation mentioned above in order to get an idea if it would be worth to set up a study of this evaluation process (and we definitely think so).

4. On the other hand collect the information in a precision adapted to practical requirements: for all we know its in general unnecessary to have the precise number of transfusions, (only 0, 1, 2-10, 11-20, $\geq$ 20),
   also the outcome grading scheme is superfluous information.

5. Less effort for long term analysis. As has been explained in section 2. mainly the first year follow up is relevant, but this as complete and in time as possible.

6. Much more information on the pretransplant situation is needed: epidemiological information, changes in the distributions on the waiting list and in the group of newly assigned patients. It is absolutely necessary to study selection effects and shifts from the general waiting list to e. g. urgency or HIT list.

So if we focus our effort on the design of a matching procedure we need in many aspects (number of variables, follow-up time) less data than are presently collected in other studies but with very high quality and more pretransplant information.

5. A proposal for the development of a matching procedure

In the previous sections we discussed how to define and compute success rate, the limitations of a matching procedure based solely on this success rate (P1), and what kind of data are needed. Now we shall describe some means to overcome the side effects of (P1) and how to derive them. Of course, the success rate is a basic ingredient of any matching procedure. But we introduce two additional parameters, the waiting time of the recipient and the probability to get a better match within the near future.

This will dampen the shortsighted greedyness of (P1) and reduce the fallback rate to exception lists.

There have been attempts (e. g.in the present Eurotransplant procedure or the recent recommendations of the 'Arbeitsgemeinschaft der Transplantationszentren in der Bundesrepublik Deutschland und in Berlin' [8]) to take these parameters into account, essentially by introducing a hierarchy in the search criteria: first look for the best HLA match, if there are equal matches, prefer the candidate with bloodgroup 0, and so on. This approach is in a certain sense inflexible, and its hard to quantify the priority of e. g. a bloodgroup 0 recipient.

We prefer to view the matching procedure as a multiattributive decision process. Let us fix some notation:

W(r)   waiting time of the recipient, time since first entering a dialysis program

S(r,d) $P(T_g(R,D) \geq 1 \mid R = r, D = d)$ = success rate for a donor/recipient combination with parameters r,d.

C(r,d)   chances of a recipient with parameters r being offered a donor kidney with parameters d to get a better match within the next year, depends on HLA typing, bloodgroup and percentage of reactive serum against random panel. (for an outline of the computation see Appendix C)

The decision process is then described by a utility function $M: \mathbb{R}^3 \rightarrow \mathbb{R}$:

Given a donor D with the specific parameters d, find a recipient R on the writing list with parameters $r_{opt.}$ such that

$$(P2) \qquad M(W(r_{opt}), S(r_{opt.},d), C(r_{opt.}, d)) = \max_{R \in WL} M(W(r),S(r,d),C(r,d))$$

There are systematic ways to assess this decision function , see e.g . Keeney, Raiffa [12], Keeney [11]. A short outline of the basic theorem and a sketch of an alternative way to assess M by discriminant analysis is described in the Appendix D.

The model is quite flexible in handling new attributes or modifications of the proposed ones:

• If felt necessary there is no problem to add a real cost term (in money)

• We might wish to treat the transplant number explicitly as an independent attribute, instead of implicitly as a parameter for the success estimate.

• One might want to introduce a penalty term for the rejection of an offered kidney.

    And so on..

The important point is that there is an essentially algorithmic way to assess the functions M, W, S, C for the procedure (P2). With the proper data collection described in section 3 we can continuously adapt the estimators S and C to the changing reality. Also the assessment of M can be institutionalized.

In any case, in order to get some idea in advance on longterm effects of changes in the matching procedure it is useful to build a simulation model of a matching system. Work on this problem is in progress (see Wujciak [15]). Again what we are missing here to draw sound conclusions is information on the pretransplant situation. We need not only posttransplant follow-up but also data on the patient pool on dialysis, the selection of patients for the waiting list, and the waiting list itself.

6. System aspects

For any proposed matching procedure computer assistance has to be offered since the size of the waiting list and the complexity of the process don't allow a decision by human inspection. Nevertheless, the final decision still has to be made by the physician and not by the system. The intention can only be to support the decision, not to dictate it. Any attempt to dictate will result in a two stage procedure: First, try to find a reasonable match for an available donor in local environment of the transplant center, and second, only if you really can't use it, offer it to the central system. This should be avoided by all means: at any point in time during the decision process the physician in charge should be aware of the best local as well as the best external candidates.

So the system should, after entering the donor information, show the ten best (in the sense of P2) external and the five best local recipients with the data on the scores W, S, C and M(W, S, C). Using colors and highlighting the system will point to specifically interesting candidates. On request additional information on the recipients (HLA-typing, bloodtransfusions, disease, ...) will be displayed. Furthermore, there should be an explanatory component being able to answer to questions like 'Why is recipient A so much preferred to recipient B?', and so on. For all of this there are tools and techniques available and we don't want to dwell on the details.

Since a general procedure will never be able to handle any special case it is useful to allow each center to specify by own criteria a small urgency list. This local list will be displayed on request during each matching process and can then be compared with the other proposed candidates. Of course, here the notion of locality is not restricted to include only a single transplant center. It may well consist of a group of nearby centers (see Grosse-Wilde et al. [8]). By this we introduce a high degree of flexibility in the system.

In order to support this type of matching procedure it is essential that the user has during the decision process transparent access to the global waiting list and his own local patient pool as well. He must be able to define his private 'urgency' or 'special cases' list in complete privacy, without any possibility of central control. The only way to give a user this security of complete privacy is to let him specify his list on his local computer only and either access the central waiting list using concepts of distributed data bases or send out over a computer network the actual lists and the updates to the local nodes of the network. Both solutions are feasible, the choice depends mainly on the underlying computer network and the offered transaction speed and network reliability: a distributed data base concept will probably work only in a fast and very reliable network.

Finally, let us point to a system design aspect which still needs further basic research. During the matching procedure the physician might want to look at very specific data of the recipient in the transplant center's data base. Also a link to the dialysis centers is desirable e.g. since the crossmatch test is unreliable for about ten days after a transfusion, this information should automatically find it's way from the dialysis center data base to the waiting list. Questions of this type will be studied in the further progress of TRAINS (see the paper of Keppel in this proceedings volume).

APPENDIX

A. Computation of graft survival

The most used method for estimation of survival in clinical trials with censored data (i.e. patients are lost to follow up) is the Kaplan-Meier estimator (see e.g. Cox, Oates [5]). Basic assumption of this procedure is that the censoring mechanism is independent of the outcome. But this is in general not true for multi-center studies, especially if concerned with long-time follow-up.

In multi-center studies the data of a certain treatment e.g. transplantation is typically collected in regular time intervals at the participating centers. The information is then sent to a central data base. There a specific patient may be lost to follow-up (LF) for two reasons:

1. The patient lost contact to the center, and this information is in the central data base.

2. According to the last report the patient is still in the study but there is no further information in central data base.

The percentage of LF in the different categories varies individually with the centers: excellent centers have many patients from abroad, hence many LF's in the first category. On the other hand, worse centers have often a sloppier reporting schedule, hence more patients in the second category.

A simple correction is the following: Let p_{ij} be a selection of patients from $i = 1, ..., m$ centers and $j = 1,...,n_i$ patients, then let $\mu_i(t)$, $\sigma_i^2(t)$ be mean and variance of percentage of 'survivals' in center i, based on Kaplan-Meier algorithm applied to $p_{i1},...,p_{i,ni}$ and then take the weighted average

$$\mu = \sum \mu_i n_i \ / \ (\sum n_i) \qquad \sigma^2 = \sum n_i^2 \sigma^2 = (n_i^2 \sigma_i^2) \ / \ (\sum n_i)^2$$

as estimate for the whole population. Another problem is that 'bad news travel faster': E.g., asked to report if a certain transplant still functions two years after transplantation this can be answered even before two years if the patient died after 15 months but in case the patient is still healthy it takes more than two years for the answer to arrive in the central DB. In general, patients having problems will be in closer contact to the center. The methods to reduce the bias introduced by this decency between outcome and time delay of data entering the central DB strongly depend on the data collection procedure. The choice of the collection procedure will be discussed by Roebruck in this volume.

B. Estimation of S(r,d)

Our problem can be described more generally in the following framework: Let $V_j, j = 1, ..., n$ be finitely valued random variables with values $v_{j i_j}$ and let S be a binary random variable then find a function p such that

$$p(v_{1 i_1}, ..., v_{n i_n}) \cong P(S = 1 \mid V = v_{1 i_1}, \ldots).$$

For short we set $I = (i_1, ... i_n)$, $v_I = (v_{i_1}, ..., v_{i_n})$, and $V = (V_1, ..., V_n)$ If the variables V_j are independent in the whole population, as well as in the subpopulation with $S = 1$, then

$$P(S = 1 \mid V = v_I) = (\prod_{j=1}^{n} P(S = 1 \mid V_j = v_{j i_j})) \cdot P(S = 1)^{-n+1}$$

and these univariate conditional probabilities could be estimated by standard techniques e.g. in our case by Kaplan - Meier method.

Of course, these indepency assumptions are not fulfilled, hence we have to look for more elaborate models. Nevertheless, this is essentially the reasoning behind the celebrated expert systems approach (see Buchanan [3], Chapter 12).

Let us introduce the notation

$$m_{0I} = \#\{\text{patients with } S = 0, V = v_I\}$$

$$m_{1I} = \#\{\text{patients with } S = 1, V = v_I\}$$

then we are interested in

$$\frac{p(v_i)}{1 - p(v_i)} \cong \frac{m_{1i}}{m_{0i}}$$

The logit models make a variance analysis like ansatz

$$\log \frac{m_{1I}}{m_{0I}} = u_0 + u_{1 i_1} u_{2 i_2} + u_{12 i_1 i_2} + ...$$

where in our situation the quotient m_{1I}/m_{0I} would be derived from the Kaplan-Meier estimate. For a detailed discussion on the interpretation of the parameters and efficient computational procedures see Fienberg [7], Bishop et al. [2]. Our estimation function p has the form

$$p(v_I) = \frac{1}{1 + e^{-(u_0 + u_{1 i 1} + ...)}}$$

This method seems to us extremely well adapted to the estimation of S(r,d). Let us shortly indicate its relation to the celebrated Cox regression.

If T is a non negative random variable with differentiable distribution $F(t) = P(T \geq t)$ Then set: $f(t) = - F'(t)$ and

$$h(t) = \lim_{\delta \to 0} \frac{P(t \le T < t + \delta \mid t \le T)}{\delta} = \frac{f(t)}{F(t)}.$$

h(t) is called the hazard function. We have

$$h(t) = -\frac{d \log F(t)}{dt} \quad , F(0) = 1$$

hence

$$F(t) = \exp\left(-\int_0^t h(s)\, ds\right)$$

The proportional hazards model now assumes that the hazard depends on a multi - dimensional variable z such that

$$h(t;z) = g(z, \gamma)h_0(t)$$

where the function g, parameterized by a vector γ describes the class of survival distributions. Let us now take $h_0 = 1$ and $g(z, \gamma) = log(1 + e^{\gamma z})$ and hence for t = 1

$$P(T \ge 1 \mid z) = F(1;z) = \frac{1}{1 + \exp^{(\gamma z)}}$$

which means that there is a direct translation of the logit model into the proportional hazards model. Nevertheless, it is an open question how the estimated parameters derived by logit models compare to those derived by Cox regression. A problem with the proportional hazards model are the assumptions on the shape of the survival distributions: when trying to estimate S(r,d) there is no reason why we should be interested to fit the whole distribution instead of just one point. Also the proportionality assumption is critical for transplant data, because the long term effect of a certain parameter combination might differ from the short term effect, so the coefficients in our model, especially γ are not constant

C. Computation of C(d,r)

Assume that the recipient has the HLA type a_i, a_j, b_i, b_j, dr_i, dr_j and let the HLA loci be independent. We now choose at random a donor with HLA types a_k, a_l, ... If MISMA rsp. MISMB MISMD denote the number of mismatches at HLA locus A rsp. B, DR then (see Barnes, Miettinien [1])

$$P(MISMA = 0) = P(a_k \in \{a_i, a_j\} \wedge a_l \in \{a_i, a_j\}) = (P(\{a_i, a_j\}))^2$$

and e.g.

$$P(MISMA = 1 \wedge MISMB = 0 \wedge MISMDR = 0)$$

$$= P(MISMA = 1)\, P(MISMB = 0)\, P(MISMDR = 0)$$

$$= [2P(\{a_i, a_j\})(1 - P(\{a_i, a_j\})) + \sum_{k \ne i,j} P(\{a_k\})^2]\, P(\{b_i, b_j\})^2 P(\{dr_i, dr_j\})^2$$

and so on for the other mismatch combinations. By evaluation of S(d,r) we can easily derive a priority list which mismatch combination are preferred to others. So for a given donor /recipient combination we can list the better matching combinations and by the above procedure compute

P(random offer gives better match than present d, r combination) = PBM (d,r)

Furthermore , let

LR(r) := probability of positive crossmatch against random panel (latest reactive serum, see e.g. Collaborative Transplant Study)

BL(r) := probability, that a random donor has compatible blood group

N := number of expected offers in a fixed time, e.g. one year

then

C(d,r) := $1 - (1-PBM(d,r) \bullet LR(r) \bullet BL(r))^N$

gives the probability to get a better acceptable offer in the next year.It's easy to generalize the computation in order to incorporate correlations between the loci.

D. The utility function M

In our decision process we try to follow different, conflicting objectives: optimal success rate, small chances to get a better offer, short waiting time. These objectives are measured by scalar variables (attributes in the nomenclature of Keeney, Raiffa [11]) S(d,r), C(d,r), W(r). In the head of the expert (in this case the physician and immunologist) exists a certain preference structure $(S,C,W) \gtrsim (S',C',W')$ saying which combination of attributes is preferred to another. We are looking for a utility function $M:\mathbb{R}^3 \rightarrow \mathbb{R}$ such that

$M(S,C,W) \geq M(S',C',W') \Leftrightarrow (S,C,W) \gtrsim (S',C',W')$

Definition: Let X,Y be attributes, $\gtrsim$ a preference structure

1. X,Y preferentially independent if $(\exists x:(y,x) \gtrsim (y',x)) \Rightarrow (\forall x:(y,x) \gtrsim (y',x))$

2. X,Y utility independent if for two lotteries (distributions) μ_y, $\mu_{y'}$

   $(\exists x (\mu_y,x) \gtrsim (\mu_{y'},x)) \Rightarrow (\forall x:(\mu_y, x) \gtrsim (\mu_y, x))$

Theorem (Keeney, Raiffa): Let $X_1, ..., X_n$ be attributes with a preference structure $\gtrsim$, and let x_i be utility independent from the complementary set of variables for all i then there exists a utility function of the form

$$u(x_1, - , x_n) = \sum_{i=1}^{n} k_i u_i(x_i) + \sum_{i=1}^{n} \sum_{j>i} k_{ij} u_i(x_i) u_j(x_j) + ... + k_{12...n} u_1(x_1) ... u_n(x_n)$$

Details and computational procedures for the assessment of the scalars k and the one - dimensional functions u_i can be found in the book of Keeney, Raiffa. The problem is that the assumptions on the preference structure, especially utility indepence are hard to verify. Also, for the derivation of the utility function all decision alternatives are given equal weight, even if many will never show up in reality.

Another approach is the following: Propose alternatives (S,C,W,), (S′,C′,W′) based on realistic assumptions on the distributions of the parameters d, r. and ask physicians or transplant coordinators for a decision. Then use discriminant analysis to find a function seperating the chosen from the rejected. The advantage of this approach is that no assumptions over the independence structure are made. Furthermore, because the sample decision are generated by a sample generated from realistic distributions of the parameters, the discriminant analysis will put most effort on separating the more probable decision alternatives whereas the systematic approach of Keeney, Raiffa goes systematically over the whole decision space, not being able to distinguish between probable and improbable decision situations. A systematic comparison of these approaches should be worthwhile.

And again, a good simulation model will be helpful to compare the effect of different utility functions.

E. Optimal stopping of renewal processes

A different mathematically attractive approach has recently been described by David and Yedical [6]: Let $0 = < t_0 < t_1 < t_2 < ...$ be arrival times of offers X_j, independent identically distributed and positive, bounded random variables with distribution function $F(x) = P(X \le x)$. Furthermore, let $\beta : \mathbb{R}^+ \to \mathbb{R}^+$ be a continuous, nonincreasing discount function with $\beta(0) = 1$. A decision at time t_j has to be made to reject or accept the offer. If the offer is accepted then we gain $\beta(t_j)X_j$. Let $G(t) = P(T \le t)$, where T denotes the survival time of the recipient, and let t_j be a renewal process with distribution function $H(s) = P(t_{j+1} - t_j \le s)$ for $j \ge 0$. We define:

$V(t,x)$ = optimal expected discounted reward if not stopped or self-terminated at t and offer x arrives.

Then

$$V(t,x) = max\{\beta(t)x, \lambda(t)\}$$

$$\lambda(t) = \int_{s=0}^{\infty} \overline{G}(s \mid t) \int_{s=0}^{\infty} V(t + s, y) \, df(y) dH(s)$$

where $\overline{G}(s \mid t) = P(T > t + s \mid T \ge t)$. λ gives 'control limit', i.e. the future expected reward if the offer is rejected.

Theorem(David,Yedical): If G is an increasing failure rate, i.e. for any $s \ge 0$ $G(s,t)$ is nonincreasing with respect to t, then there exists an optimal policy characterized by a continuous, nonincreasing function $\lambda : \mathbb{R}^+ \to \mathbb{R}^+$ such that an offer is accepted if $\beta(t)x \ge \lambda(t)$.

Before this theory can be applied to our problem, it has to be generalized to an n-person assignment problem: There is more than one recipient waiting on the list. And even if the theory can be generalized it is far from clear that a numerically feasible solution exists. Besides this, it is hard to get empirical data, e.g. for the discount factor of the recipients.

F. A remark on patient selection

In this section we take the example of the selection hypothesis for the transfusion effect to exemplify that certain treatments and their beneficial effect might be viewed different by the physician and the patient. The selection hypothesis (see Mickey [13]) says that pretransplant transfusions tend to sensitize those patients (hence raise their chances of a positive crossmatch, thus reducing their chances to get a donor kidney) which tend to reject foreign tissue with higher probability, anyway. So it may be viewed as a test: with transfusions the chances to pass the crossmatch test are worse, but if passed there is a higher successrate. This is described (with artificial numbers, since we don't have the data to study patient selection), by the decision tree in figure 2.

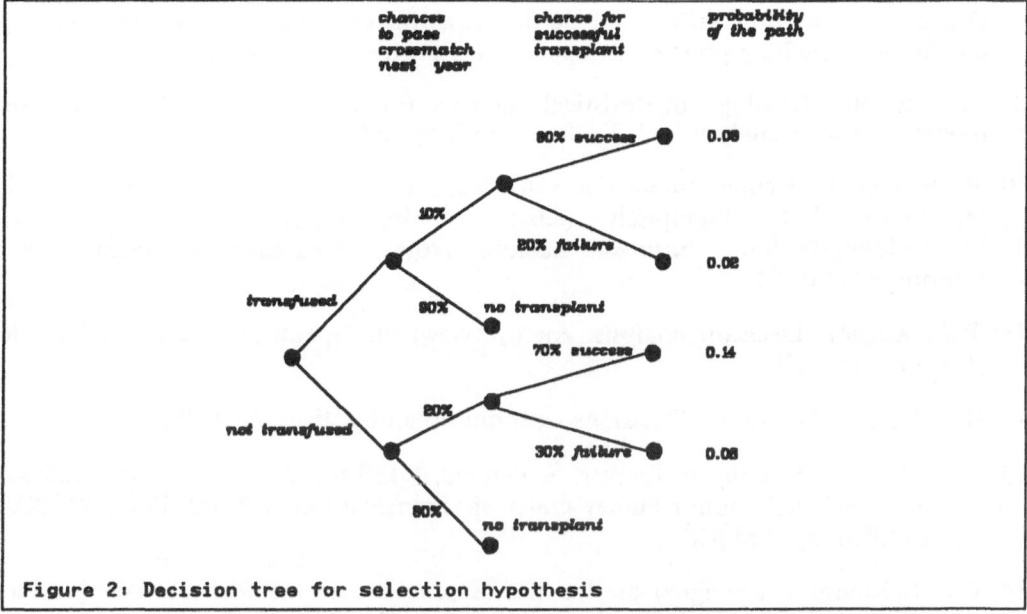

Figure 2: Decision tree for selection hypothesis

The physician would probably prefer the upper half of the tree: the relative failure rate under the transplanted patients is smaller. The patient might on the other hand prefer the lower half: he might accept the higher relative risk of failure for higher absolute chances to get a successful transplant. But as long as there are no precise data on the pretransplant situation all the examples and studies are artificial and not decisive. A clearer knowledge on pretransplant patient selection is required for a better understanding of this and many other possible side effects of matching procedures.

References

1. B.A.Barnes, O.S. Miettinnen: The search for an HLA and ABO Compatible cadaver organ transplantation; in: Transplantation Vol.13, No 6 (1972) pp. 592-598

2. Y.M.M. Bishop, S.E. Fienberg, P.W. Holland: Discrete multivariate analysis; MIT Press (1975)

3. B.G. Buchanan: Rule based expert systems; Addison Wesley (1984)

4. J.Cicciarelli, M.R. Mickey, P.I. Terasaki: Center effect and kidney graft survival; in: Transplantation Proceedings, Vol. XVII, No. 6 (1985), pp. 2803-2807

5. D.R. Cox, D. Oakes: Analysis of survival data; Chapman & Hall (1984)

6. I. David, U. Yedical: A time-dependent stopping problem with application to live organ transplants; in: Operations Research Vol. 33, Wo 3 (1985) pp. 491-504:

7. S.E. Fienberg: The analysis of cross-classified categorial data; MIT Press (1980)

8. H. Grosse-Wilde, G. Opelz: Recommendations of 'Arbeitsgemeinschaft f}r Organtransplantation' (FRG and Berlin West) for crossmatch, antibody screening and kidney matching policies; to appear in: Eurotransplant Newsletter

9. L. Guttman: The illogic of statistical inference for science; in: Applied stochastic models and data analysis Vol. 1, No 1 (1985) pp. 3-9

10. R. Janßen, R. Reuter: Interactive statistical analysis of transplant data; in: H.J. Jeschinsky, H.J. Trampisch (eds.): Medizinische Prognose - und Entscheidungsfindung, Springer Lecture Notes 'Medizinische Statistik und Informatik' Vol. 62

11. R.L. Keeney: Decision analysis: An overview; in: Operations Research Vol. 30 (1982) pp. 803-838

12. R.L. Keeney, H. Raiffa: Decisions with multiple objectives; John Wiley (1976)

13. M.R. Mickey, S. Cats, B. Graver, S. Perclue, P.I. Terasaki: Transfusion and selection in cadaveric donor kidney grafts; in Transplantation Proceedings, Vol.XV, No. 1 (1983), pp. 965-968

14. D.S. Salsburg: The religion of statistics as practiced in medical journals; in: The American Statistician Vol. 39, No. 3 (1985), pp. 220-223

15. Th. Wujciak: A Simulation environment for donor/recipient matching; Diploma thesis, in preparation

COMPUTER AIDED DECISION SUPPORT FOR POST-TRANSPLANT RENAL ALLOGRAFT REJECTION

Hans-Georg Müller[1,3], Thomas Müller[2], Christian Lohrengel[1], Harald Lange[2]

[1] Institut für med.-biol. Statistik, Ernst-Giller-Str. 20, 3550 Marburg, F.R.G.

[2] Abtl. Nephrologie, Klinikum Lahnberge, 3550 Marburg, F.R.G.

[3] Division of Statistics, Univ. of California, Davis, Ca. 95616, USA

Table of contents:

0. ABSTRACT

Following renal transplantation a patient has to be monitored closely in order to detect the onset of a rejection episode as soon as possible. The decision whether to enhance immuno-suppressive therapy - with its severe side effects - to suppress the rejection episode is often difficult, especially for the less experienced clinician. Therefore, objective methods of decision support are needed. The decision is based on observing the time course of clinical parameters like serum creatinine, urea, urine volume, urinary sodium, osmolarity, and body weight.

Our contribution deals with three questions:
(1) What is the distribution of rejection episodes over time elapsed since transplantation?
(2) How can current values of clinical parameters be predicted from past observations and the discrepancies between prediction and observation be used for the diagnosis of an episode?
(3) Which of the variables have the highest diagnostic value?

For (1) we use a nonparametric kernel method with automatic bandwidth choice by cross-validation. For (2) we developed a nonparametric kernel regression estimator with one-sided kernels. For (3) we employ logistic regression and other multivariate discriminant methods. Future prospects are to implement the developed FORTRAN programs on a microcomputer in order to supply online decision support.

1. INTRODUCTION

Renal transplantation today has become an established and justified mode of treatment for patients in end-stage renal failure. Human renal transplantation was made possible due to three main factors:
1. the knowledge of immunology and tissue typing, which allows selection of the most histocompatible donor;
2. the use of immunosuppressive therapy for the suppression of the recipient's immune response; and
3. the availability of effective replacement therapy for the waiting patient of following graft loss.

Great advances, mainly the additional HLA-DR matching (1-4), the use of pre-transplant transfusions (5,6) and the institution of the new immunosuppressive agent ciclosporine (7,8), have led to a most striking decline in patient morbidity and mortality (9,10). The improvement of graft survival rates has been impressive as well, with an overall 20 to 30% increase in the one-year survival rate (11,12).

But the major cause for graft loss remains allograft rejection (13,14). Therefore optimizing immunologic compatibility with organ sharing and improvement of anti-rejection therapy is necessary.

For effective and successful therapy early and reliable diagnosis of graft rejection is mandatory. A rejection process causes a deterioration of graft function resulting in a decline of the glomerular filtration rate (GFR) and, as the filtration rate is reversely related to serum-creatinine (s-crea), an increase in the s-crea level. S-crea seems to be the parameter with the greatest reliability and availability in the detection of renal functional impairment (15). For monitoring the s-crea reciprocal is recommended as a more suitable indicator than the s-crea itself (16,17,18). Its graphical presentation in the postoperative course using computers and statistical aids helps clinical decision making, in addition it reduces the subjective component in the diagnosis of rejection and allows standardization and comparability (19,20).

According to current knowledge and corresponding to clinical practice, the diagnosis of a rejection episode can not be based only on one clinical parameter. Also it is doubtful whether there is a clear, uniquely defined set of patterns of the different parameters that would be always associated with a rejection episode. An approach to detect abrupt changes in pattern of s-crea and of s-urea was developed in a series of papers by A.F.M. SMITH and others (21,22,23) using a modified Kalman filter technique. However, "slow" rejection episodes can also occur when there are initially only minor changes in the parameters.

In our approach we include from the beginning three clinical parameters (s-crea, urea, urine volume). Assuming that these parameters follow a smooth time course, we apply a nonparametric kernel estimation technique to estimate the values of the time course of each of these parameters at the current day where the decision is to be made from past observations. We also estimate the first and second derivative (slope and curvature). These quantities as well as the corresponding quantities from the two preceding days and also the observed parameter values of these same days serve as input variables for discrimination between current days with and without rejection. The decision whether a rejection occured at a given current day, needed as an objective external information, was made retrospectively by an experienced nephrologist having access to all clinical curves and data concerning the patient.

This approach of discrimination of time courses is completely data-adaptive. Stepwise logistic discrimination was used to filter out variables of predictive quality from observations, reciprocal observations, estimated time courses and first two derivatives of observations and of reciprocal observations for the three parameters on three days. The apparent correct classification rate of this model was about 95% in both groups with and without beginning rejection episode; the cross-validation correct classification rate was 85%. The resulting classifier was programmed and can be used for on-line decision support. For each day, the probability of a rejection episode is computed prospectively; a cut-off point can be specified.

The paper is organized as follows: Section 2 contains a discussion of clinical features of rejection episodes and describes current practice of clinical decision making. In Section 3, the kernel method is presented for estimating the probability density of the occurence times of rejection, and in another version as a tool for the prediction of current calues of a time course as well as of its first two derivatives from past observations. The application to clinical data and the resulting logistic discrimination function is decribed in Section 4. Section 5 contains a complete listing of FORTRAN computer source code which can be implemented for on-line decision support.

2. ALLOGRAFT REJECTION IN RENAL TRANSPLANTATION

2.1. POSTOPERATIVE CLINICAL COURSE

The postoperative progress of transplant function shows a high degree of variability, rejection is only one possible cause for functional deterioration.

Ideally there is immediate postoperative transplant function, the typical pattern for living related donor transplants. The initial polyuria returns to a normal value after 24 to 48 hours, the s-crea falls to levels well below 4 mg% by the 3rd day, dialysis therapy is not necessary (22).

Immediate or early graft nonfunction is usually due to nonimmunologic lesions, i.e. premortem incidents, harvesting procedure, preservation methods. It is common in cadaver kidneys and its incidence varies from 28% to 74% (23,24). The pathological hallmark is acute tubular necrosis, which is reversible (25). These periods of oliguria may last as long as 3 to 6 weeks, the difficulty is to detect superimposed acute rejection processes which should be treated.

Delayed acute renal failure episode may be due to immunologic rejection episodes as well as non-immunologic causes. They may occur at any time, but the highest incidence of rejection episodes is found during the first 6 months, especially during the first weeks. 85% of delayed renal failure episodes occur during the first month (26). The early diagnosis of these acute rejections is necessary as prompt treatment may prevent irreversible damage. The slow progressive nonepisodic late failure of transplants with a concomitant slow decrease in GFR is usually caused by chronic graft rejection and may arise following long periods of stable graft function (9).

2.2. CHARACTERIZATION OF ALLOGRAFT REJECTION

Allograft rejection is the most important factor limiting success in renal transplantation.

The rejection of a kidney graft is caused by a combined cellular and humoral immunologic response of the recipient against the donor's incompatible tissue antigens. Besides differences in the major histocompatibility complex there are further minor, non-HLA-dependent antigen systems which do play a role in the induction of rejection processes (27,28).

The immunologic response results in a progressive damage of the graft with pathological features of predominating tubulointerstitial or vascular lesions. Four main clinical types of rejection are differentiated due to their onset and course.

The irreversible hyperacute rejection caused by preformed cytotoxic antibodies and occuring within minutes of revascularization. After the institution of regular pretransplant cross-match reactions this form of rejection is very rare.

The delayed hyperacute or accelerated rejection episode is probably due to a secondary immune response of cytotoxic antibodies which are reactivated by the graft (22,14). It occurs between the 2nd and 10th postsurgical day and is usually irreversible (29,30,31).

By far the most common type of rejection is the, when treated early and vigorously usually reversible, acute rejection episode. 60% of all patients do have acute rejection episodes during the first 4 weeks (32), 80% occur in the first 4-6 weeks (14), being the most frequent during the initial 2 weeks (22), with a peak between the 4th and 9th day according to 14,a peak on the 22nd day according to 31. The histologic hallmarks are interstitial oedema and particularly perivascular mononuclear cell infiltration due to a primarily cell-mediated response of the host.

The chronic rejection results from the constant immunologic conflict between host and transplant. Humoral as well as cellular responses result in an irreversible progressive destruction of the graft.

2.3. DIAGNOSIS OF REJECTION EPISODES

The early detection of rejection episodes is necessary to start adequate therapy
to prevent permanent damage to the graft.
Despite a great variety of disease processes damaging the kidney, the organ has
only a limited number of ways to react (15). This nonspecifity of clinical manifes-
tations explains the lack of a single pathognomonic parameter for rejection indu-
ced renal damage.
An increasing s-crea level due to renal functional impairment remains the most
sensitive and accepted marker of a rejection episode. In addition a great number of
other parameters are used in the differential diagnosis and assessment of renal
functional impairment in a transplant patient.
The main parameters are the following(9,13,14,15,22,32).

1. Clinical and serum parameters including fever, tenderness and increased size of
   the graft, rise in blood pressure, weight gain, oliguria, fall of the creatinine
   clearance, increase in serum amyloid A and beta-2-microglobulin (33-35), fibronectin levels (36);
2. Urinary parameters like a decline in the urine volume, fall in osmolality, decrease in sodium concentration, proteinuria, change in electrophoretic patterns
   (37), occurrence and changes in enzyme levels, e.g. lysozyme (38-40), N-acetyl-
   β-D-glucosaminidase (41-43), aminopeptidase (44), β2-microglobulin (34,45,46),
   excretion of fibrinogen split products (47), thromboxane (48);
3. Radionuclide investigations such as different renograms and scintiscans (49);
4. Radiological examinations including angiography, ultrasound scans, pyelographies;
5. Immunologic assays as lymphocyte mediated cytotoxicity (LMC), antibody dependent
   cellmediated cytotoxicity (ADCC), complement dependent cytotoxicity (CDC), inter-
   leukin-2-lymphocyte response (see 50);
6. Transplant biopsies with histological examination under light, electrone and im-
   munofluorescent microscopy and fine needle aspiration (14,51,52,53).

Despite the vast number of different parameters in the clinical routine the detection
of functional impairment by a rise in the s-crea level, the exclusion of current
nonimmunologic causes, in doubtful cases the performance of a transplant biopsy,
and the positive response to the immunosuppressive anti-rejection therapy remain
the hallmarks for the diagnosis of an acute rejection episode.

3. COMPUTER METHODS FOR DETECTION OF POSTTRANSPLANT REJECTION EPISODES

3.1. ESTIMATION OF THE PROBABILITY DENSITY OF THE REJECTION TIME

For deciding whether a rejection episode is starting at a given current day, it is of general clinical interest to know the distribution of rejection episodes over time, where time always refers to time elapsed since transplantation. In order to avoid problems with censored data, it is assumed here that all patients have been observed and closely monitored until time $T > 0$. A second assumption is that the observed times of rejection episodes $T_i \geq 0$ are independent in the group of patients. Of course this is an idealization, since immediately after a rejection episode a second episode could not be detected. Assuming that the distribution of rejection times T_i has a probability density f, the aim is the estimation of f on the interval [0,T]. A classical approach is to draw a histogram, which however is not uniquely defined and also not smooth. Better methods of density estimation are available nowadays. An up-to-date overview is given by SILVERMAN (54).

One approach that can be recommended is the kernel method (55,56), where the estimate at $x \in [0,T]$ is defined as

$$\hat{f}(x) = \frac{1}{nb} \sum_{i=1}^{n} K(\frac{x-T_i}{b}) \quad .$$

K is the kernel function which has to satisfy $\int K(v)dv=1$, $\int K(v)vdv=0$, and b the bandwidth which regulates the trade-off between variance and bias. Choice of b is crucial for the performance of the kernel estimator, and we use an automatic method of bandwidth choice proposed in (57). The idea is to minimize the integrated square error

$$\int (\hat{f}-f)^2 = \int \hat{f}^2 - 2\int \hat{f}f + \int f^2$$

w.r. to the bandwidth b; first term on the r.h.s. can be evaluated by numerical integration, the last term does not depend on b and the middle term can be approximated by cross-validation:

$$\int \hat{f}f \approx \frac{1}{n} \sum_{i=1}^{n} \sum_{j \neq i} \frac{1}{b} K(\frac{x_i - x_j}{b}) \quad .$$

Another problem concerns choice of the kernel function K. Usually the function $K \equiv \frac{3}{4}(1-x^2)1_{[-1,1]}$ can be recommended, but a specific problem occurs near the end points 0 and T; since there $[x-b, x+b] \not\subset [0,T]$. This problem can be resolved by the use of boundary kernels satisfying the moment conditions above but constructed with asymmetric supports [-q,1] (right boundary) resp. [-1,q] (left boundary), where $q = (1-x)/b$ resp. $q = x/b$. More details can be found in (58). A FORTRAN subroutine for density estimation including automatic bandwidth choice and boundary modification was written by the authors and a listing can be obtained from C. Lohrengel.

3.2 KERNEL METHOD FOR PROGNOSIS

In contrast to 21,19,18 we assume that the true time course of the clinical parameters is smooth, but the available measurements are contaminated with errors.
These errors are caused by timing errors, variations on short time scales; laboratory, physiological or clinical measurement errors, and disturbances due to dialysis of the patients to name a few (see 21 for a discussion of such errors).
Besides the actual measurements, estimates of the true time course are sought which
only make use of past measurements. Another aim is the estimation of the first
and second derivatives which might be early indicators of level resp. slope changes.

The assumptions might be formalized as follows: The measurements y_i are obtained
during the time interval $[0,T]$ as $y_i = g(t_i) + \varepsilon_i$, $i=1,...,n$, where $0 \leq t_1 \leq t_2 \leq ... \leq t_n \leq T$
are the times of measurement (which in principle could be irregularly spaced) and
ε_i are contaminating errors. A basic assumption is that these errors are uncorre-
lated for different measurements. A kernel method for the estimation of g, $g^{(1)}$,
$g^{(2)}$ at t from the noisy data (t_i, y_i), $i=1,...,n$ was developed in 59:

$$g^{(\nu)}(t) = \frac{1}{b^{\nu+1}} \sum_{i=1}^{n} \int_{s_{i-1}}^{s_i} K(\frac{t-u}{b}) du \, y_i \quad ,$$

where K is the kernel function, b the bandwidth and (s_i) satisfies $s_0 = 0$, $s_i = \frac{1}{2}(t_i + t_{i+1})$,
$i=1,...n-1$, $s_n = T$. This kernel estimate can be interpreted as a moving weighted average where
the weights are determined by K and b. The kernel function has to satisfy the moment conditions

$$\int K(v) v^j dv = \begin{cases} 0 & 0 \leq j < k \\ (-1)^{\nu} \nu! & j = \nu \\ \neq 0 & j = k \end{cases}$$

for some given number $k \geq \nu+2$, the so-called order of the kernel. Usually the support
of the kernel function is chosen to be $[-1,1]$. However, for purposes of monitoring
this is not possible, since only data from the past are available. Therefore, we
have to choose kernels with asymmetric support $[-q,1]$, where it proved after some
experimentation reasonable to choose $q = \frac{-1}{25}$. These kernels K_q can be selected as
polynomials of degree at least $(k-1)$, where the coefficients are determined by
a linear system of equations resulting from the moment conditions, see(58). It is
adviseable to require in addition $K_q(1)=0$, $K_q(-q)=0$ which leads to smoother
curve estimates. The kernel polynomials K_q are then of degree $(k+1)$. Examples of
weights given to the measurements y_i when using such kernels with t=10 and b=10
are shown in figure 1 for $\nu = 0$, k=2, in figure 2 for $\nu=1$, k=3 and in figure 3 for
$\nu=2$, k=4.

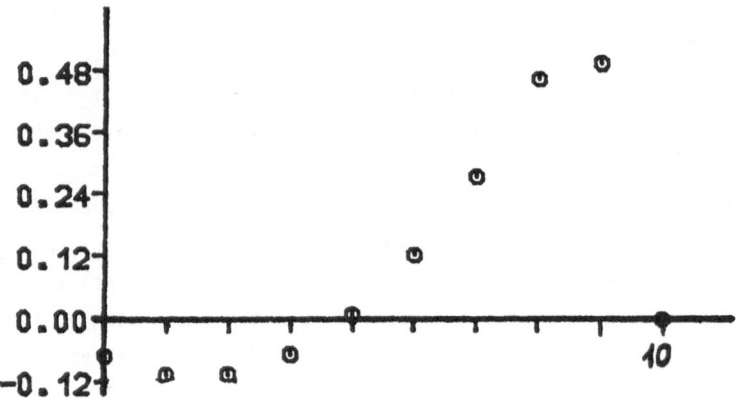

Fig. 1 Weights derived from kernel ν=0, k=2

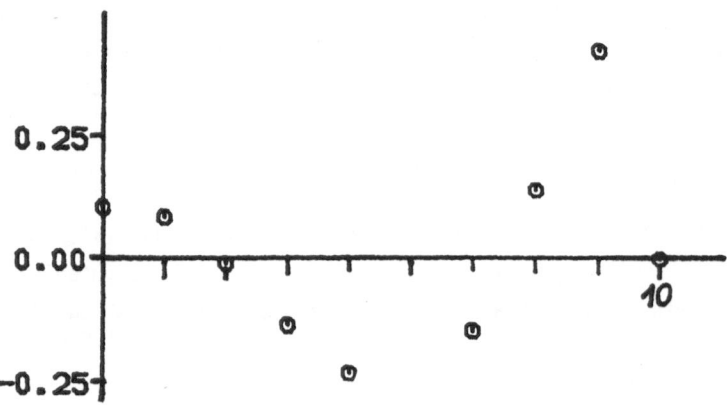

Fig. 2 Weights derived from kernel ν=1, k=3

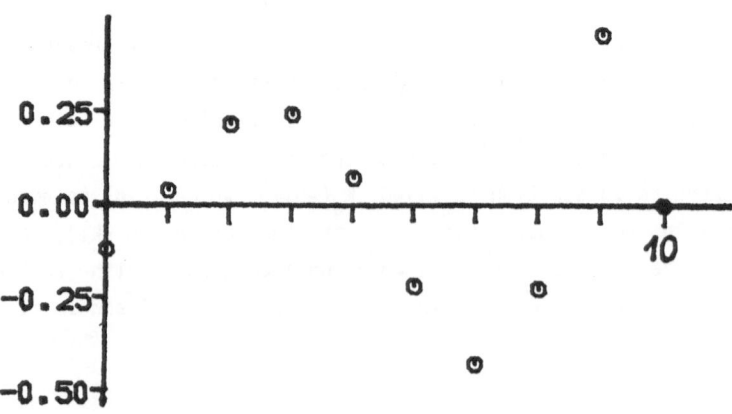

Fig. 3 Weights derived from kernel ν=2, k=4

A critical point is again the choice of the bandwidth (smoothing window). Smaller bandwidths lead to higher variance and smaller bias, i.e. changes in the time course are easier detectible with smaller bandwidth at the cost of more false positives. For the prediction of changes, the bandwidth regulates the trade-off between sensitivity and specificity. In the current version of the decision support algorithm the user has to specify a factor by which all bandwidths occurring in the program are multiplied. In this way, sensitivity and specificity of the prediction algorithm can be adapted to specific clinical circumstances. This adaptation could also be done automatically by using cross-validation, but this is a future prospect. Cross-validation (see 60) was used in order to obtain bandwidths for the initial estimates of the time courses of the parameters and their derivatives, separately for each individual, which were entered into the logistic discrimination procedure. The mean values of these bandwidths are used in the current version of the program. For $\nu > 0$ (derivatives over the individuals) the same bandwidths were employed as for $\nu = 0$. One might improve on the present algorithm by using more sophisticated methods for bandwidth choice for derivatives. Cross-validation for $\nu = 0$ aims to minimize

$$CV(b) = \sum_{i=1}^{n} (\hat{g}_{-i}(t_i) - y_i)^2$$

w.r. to b, where $\hat{g}_{-i}(t_i)$ is the kernel estimate at t_i without using observation t_i.

4. CLINICAL STUDY
4.1. CLINICAL DATA

The study was retrospectively performed on 26 patients, 7 women and 19 men, aged from 16 to 63 years (x=39.6). All patients were treated with cadaver kidney transplants (25 first and 1 second transplantation) at the university hospital of Marburg between September 1980 and February 1982. Following transplantation they were studied for a period up to 718 days (September 1982), in patients with functioning grafts the mean observation time was 389 days, in those with functional graft loss 145 days. In the postoperative course 8 grafts showed immediate function, in 17 patients dialysis therapy in the early postoperative period was necessary (x=11.5 days), and in 1 patient there was no sufficient transplant function at all. The quality of HLA-matching between donor and recipient was assessed by the "HLA-classes" according to Eurotransplant (61).
In September 1982 16 patients had a functioning graft , in 9 patients dialysis therapy had to be reinstituted and one patient had died due to an extra-renal complication (pulmonary embolism). The standard immunosuppressive regimen consisted of azathioprine and prednisone. Routinely daily measurements of serum-creatinine, urea,

urinary volume, sodium excretion, urinary osmolarity, serum and urinary lysozyme, body weight, and temperature were performed.A steroid pulse was given when a rejection episode was clinically diagnosed or suspected following the conventional clinical and laboratory parameters as described in 2.3. Altogether 74 rejection episodes (65 reversible and 9 irreversible) were considered to have occurred during the observed period on the ward or in the outpatient clinic.

For the development of the computer programs in this study 19 rejection episodes were included. All were acute, reversible rejections, occurred on ward and were therefore under continuous close observation, caused a definite, distinct and steroidpulse-sensitive deterioration of graft function, were independent of dialysis therapy and occurred in the absence of any overt non-immunologic cause for graft impairment.

4.2. DISTRIBUTION OF REJECTION EPISODES OVER TIME

The application of the automatic kernel density estimate described in section 3.1. to the times of the rejection episodes yielded the following results: The acute rejection episodes are most likely to occur around the 20th day. Analysing the occurrence of graft rejection episodes according to the immunologic compatibility between graft and recipient shows no difference between the better matched grafts (HLA class ≤ 6) in contrast to those with a poorer HLA-A,-B and -DR matching (HLA class > 6) in the early maximum around the 20th postoperative day. But interestingly the late rejections after day 60 following transplantation seem to belong all to the group with more HLA mismatches (see Fig. 4).

The density of all acute rejection episodes that were observed for more than 400 days following operation is shown by the next diagram. After the early maximum at about 25 days there is a rapid decline until day 75. After the first three postoperative months the density is more or less constant at a low value, it approaches 0 at about day 225. Nevertheless the clinician always must be alert to possible rejection processes, much later graft rejections do occur (see fig. 5).

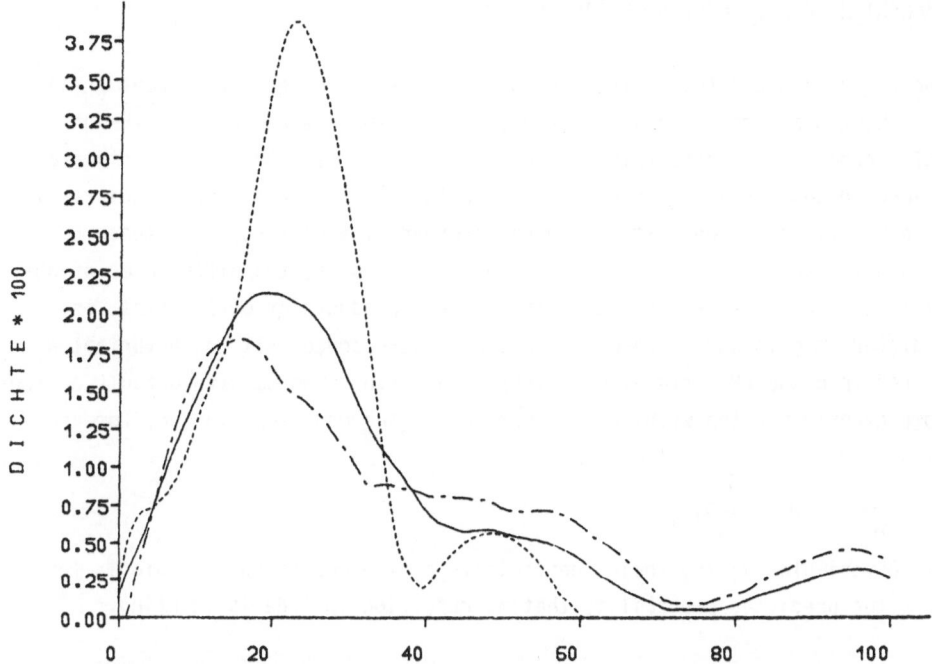

Fig. 4 Kernel-estimated density of all acute rejection episodes observed over
100 days (solid line) and of the subgroups with HLA-class $\leq$ 6 (dashed line)
(n=25) and HLA-class > 6 (dash-dot line) (n=25)

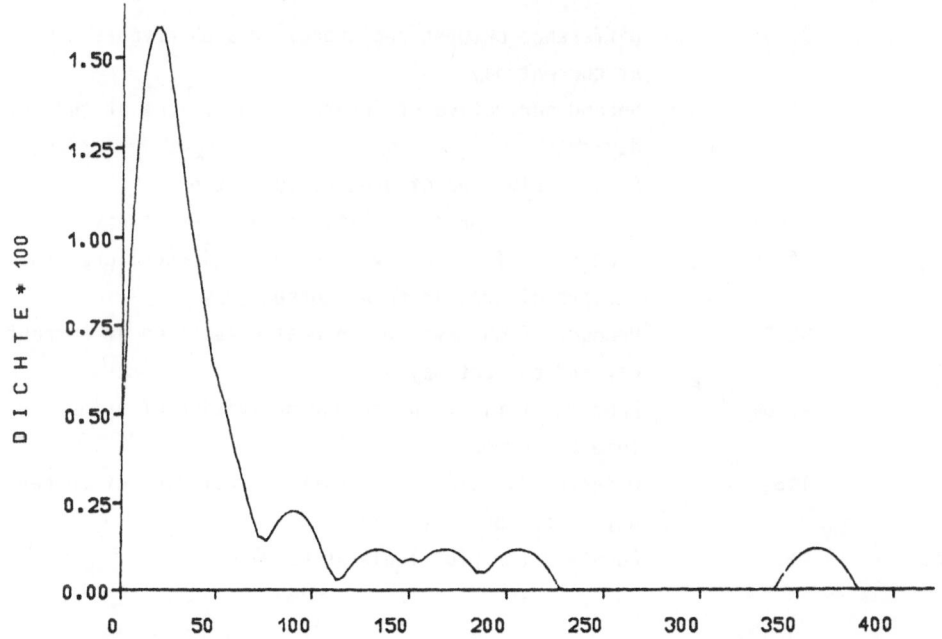

Fig. 5 Kernel estimated density of all acute rejection episodes observed for at
least 400 days (n=30)

4.3 AN ALGORITHM FOR ON-LINE REJECTION DIAGNOSIS

For selected undisturbed (by dialyses or previous rejection episodes) days (n=111)
with or without starting rejection episode, and the two preceding days, the observation of s-crea, urea, urine volume, their reciprocals, and the kernel estimates
as described in section 4.2 of s-crea, urea, urine volume and of their reciprocals
(including the first two derivatives) were recorded. The differences between observed and estimated values for all parameters at all days, the differences of observed values between successive days, and several products of derivatives, including different days and different parameters, were added. All these variables
were entered into the BMDP stepwise logistic regression program with default options.
After some experimentation with derived variables, the model obtained by forward
stepping is

$$p = \frac{e^u}{1+e^u} \quad , \quad u = \sum_{i=1}^{8} a_i v_i$$

where the 8 variables v_i and their coefficients a_i are described in Table 1. The
p value is the predicted probability that no rejection episode is starting.

Table 1 Variables in the logistic discrimination model

Variable	Coefficient	Description
IHST2D	3520	Difference between reciprocal urea observed-estimated at current day -1
IUV3A2	-11000	Second derivative of inverse urine volume at current day
HST3A1	-0.0853	First derivative of urea at current day
IKR30	-54.7	Observed reciprocal creatinine at current day
IHK3A1	-46400	Product of first derivatives of reciprocal urea and reciprocal creatinine at current day
IKR230	89.0	Product of observed reciprocal creatinine at current day and current day -1
UV3A	$-7.04 \cdot 10^{-6}$	Product of first and second derivative of urine volume at current day
IKROD	143	Observed difference of inverse creatinine at current day - (current day - 1)
Constant	11.7	Constant term for logistic model

Therefore the final algorithm requires computation of the following kernel estimates:

order of derivative	Day	Parameter	Required for
0	2	reciprocal urea	IHST2D
2	3	inverse urine volume	IUV3A2
1	3	urea	HST3A1
1	3	reciprocal creatinine	IHK3A1
1	3	reciprocal urea	IHK3A1
1	3	urine volume	UV3A
2	3	urine volume	UV3A

The corresponding bandwidths (up to a common factor, the so-called tuning parameter which has to be provided by the user) are to be found in the listing of the source code of PROGNO in the appendix. With a cutpoint of $p=.80$, 100% of the 19 rejection episodes in the data and 91% of the other days would have been classified correctly. Since this apparent correct classification rate is too optimistic, a cross-validation estimate of the correct classification rate by omitting three times one third of the cases as test sample, computing the classification function only from the remaining cases (learning sample) and predicting the cases in the test sample. This procedure yielded an average correct classification rate of 85%.

After the logistic discrimination function was determined, it was implemented in an algorithm for decision support. The source code of this algorithm is completely listed in the appendix. It consists of the subroutine PWERT which computes the p value for each day by using the coefficients of Table 1. A cutpoint of $p=0.55$ was found to be adequate, considering specificity and sensitivity. If $p<0.55$, a rejection episode might be suspected, otherwise not. PWERT is called by the subroutine PROGNO which evaluates the logistic discrimination function. A critical input parameter for PROGNO is the tuning parameter which is a factor by which all bandwidths of the program are multiplied. It has to be individually adjusted by checking sensitivity and specificity of the decision support algorithm. For our data, the factor 1 was reasonable. PROGNO calls KERN4, the core subroutine which performs kernel estimation and differentiation. KERN4 requires the subroutines KEFFBO for the computation of the kernel polynomial and KEWEIB for the computation of the kernel weights. KEFFBO depends on the subroutine KERSOL for the solution of the linear system of equations resulting from the moment conditions on the kernel for the coefficients of the kernel polynomial.

Summarizing, the structure of the program is visualized in fig. 6.

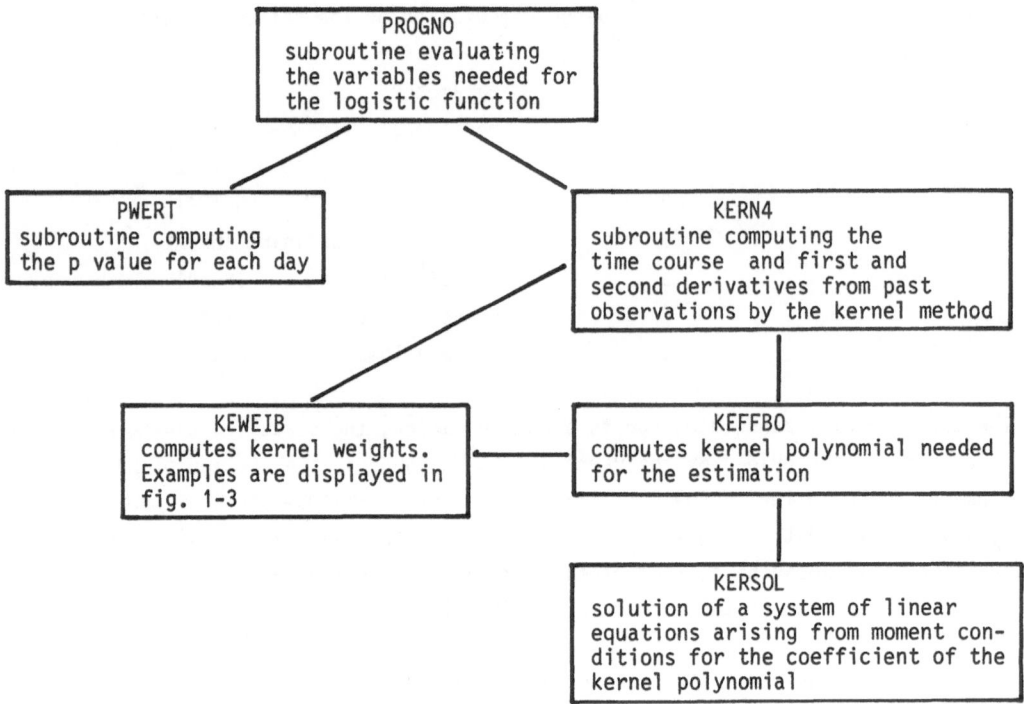

<u>Fig. 6</u> Structure of the program for on-line decision support

The input to PROGNO consists of the three series of s-crea, urea, urine-volume measured daily from the first postoperative day up to the current day and of a tuning factor for the bandwidths which regulates sensitivity and specificity of the procedure; (the default value is 1.0).

Since past observations are necessary to compute the kernel estimates, the present version of the algorithm computes p values from postoperative day 8 on; the first seven days can not yet be monitored. The kernel algorithm is for large numbers of measurements also consistent if the input data are observed at irregular time intervals. In simulations that were carried out, the results with irregular observations however were much worse than with regular observations.

Examples for the application of this algorithm are shown in fig. 7+8

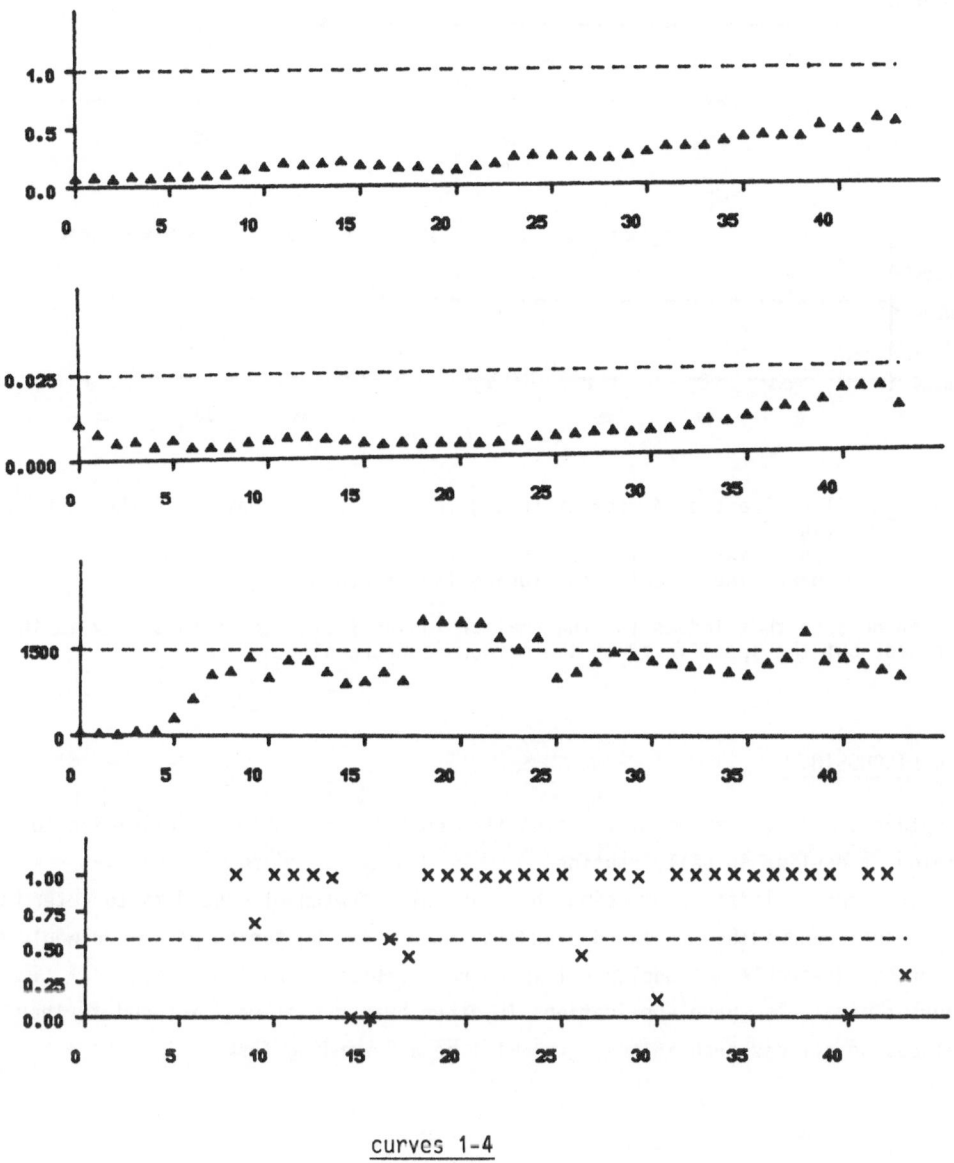

curves 1-4

Figure 7: Time courses for one patient over 40 days after transplantation

curve 1: observed time course of inverse creatinine
curve 2: observed time course of inverse urea
curve 3: observed time course of urine volume
curve 4: p value computed by the program, using tuning factor 1.0
A p value falling below the broken line suggests the onset of a rejection reaction

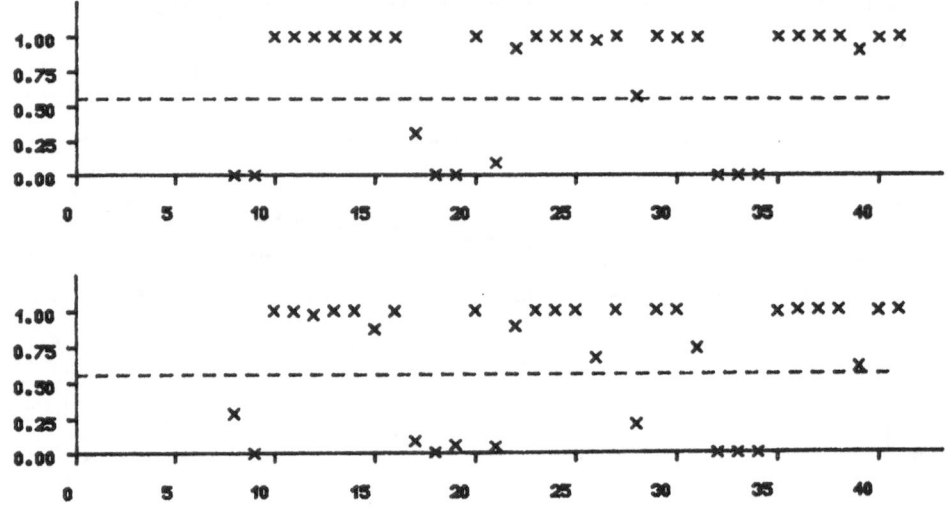

Figure 8: The effect of different tuning factors for one patient (other than in fig. 7)
upper line: p value for tuning factor 1.0
lower line: p value for tuning factor 0.875

It can be seen that indeed for the smaller tuning factor the sensitivity is increased, but the specificity of the method is decreasing.

4.4 DISCUSSION

A future aim is to implement the algorithm on PC's to enable the clinician to employ it on-line in daily routine.
On-line cross-validation by using the available stretch of past data to determine individual bandwidths at each day and to get rid of the tuning parameter would be a further desirable methodological improvement. Discrimination of clinical time courses might also have applications in other branches of medicine and further methodological research in this area will be a rewarding task.

5. REFERENCES

1. Persijn GG, Gabb BW, van Leeuwen A, Nagetagaal A, Hoogeboom J, van Rood JJ: Matching for HLA antigens of A,B, and DR loci in renal transplantation by Eurotransplant. Lancet 1978; 1:1278-81.
2. Ting A, Morris PJ: Matching for B-cell antigens of the HLA-DR series in cadaver renal transplantation. Lancet 1978,1: 575-77.
3. Albrechtsen D, Arnesen E, Solheim BG, Thorsby E: Significance of HLA-DR matching and of B cell cross-match texts in vitro and in cadaver renal transplantation. Transplant.Proc. 1979; 9:743.
4. Albrechtson D, Flatmarck A, Jervell J: HLA-DR matching in cadaver renal transplantation. Lancet 1978; 1:825
5. Opelz G, Sengar DPS, Mickey MR, Terasaki PI: Effect of blood transfusions on subsequent kidney transplants. Transplant. Proc. 1973; 5:253-59.
6. Opelz G, Terasaki PI: Improvement of kidney graft survival with increased numbers of blood transfusions. New Engl J Med 1980; 299: 799-803.
7. Calne RY: Immunosuppression for organ grafting. Observations on cyclosporin A. Immunol. Rev. 1979; 46:113.
8. European Multicenter Trial Group: Cyclosporin in cadaveric renal transplantation: one-year follow-up of a multicentre trial. Lancet 1983; 2:986.
9. Strom TB, Tilney NL: Renal transplantation: Clinical Aspects. In: Brenner BM, Rector FC (eds.): The Kidney. W.B. Saunders Company, Philadelphia, 1986, p. 1941.
10. Carpenter CB, Marrill JP: Histocompatibility and transplantation. In: Harrisons Principles of Internal Medicine. Mc Graw-Hill Internat. Book Company, ninth edition, 1980, pp. 360.
11. Brunner FP, Brynger H, Chantler C, Douckerwolcke RA, Hathway RA, Jacobs C, Selwood MH, Wing AJ: Combined report on regular dialysis and transplantation in Europe IX 1978. Proc. Eur. Dial. Transplant Assoc. 1979; 16:4.
12. Opelz G: Collaborative Transplant Study. CTS Newsletter. 1986;1.
13. Carpenter CB: The early diagnosis of renal allograft rejection in man. Advances in Nephrology from the Necker Hospital (Chicago) 1975; 5: 229-56.
14. Huland H, Klosterhalfen H: Pathomechanismus, Morphologie, Diagnostik und Therapie der Abstoßung transplantierter Nieren. Urologe A 1982; 21: 265-73.
15. Maher JF: A logical approach to the diagnosis of renal transplant rejection. Immunologic, ischemic and inflammatory impairment of renal function. Am J Med 1974, 56: 275.
16. Mitch WE,Walser M, Buffington GA, Lemann J: A simple method of estimating progression of chronic renal failure. The Lancet 1976,: 1326-28.
17. Rutherford WE, Blondin J, Miller JP, Greennalt AS, Vavra JD: Chronic progressive renal disease: rate of change of serum creatinine concentration. Kidney Intern. 1977, 11:62-70.
18. Knapp MS, Smith AFM, Trimble IM, Pownall R, Gordon K: Mathematical and statistical aids to evaluate data from renal patients. Kidney International 1983, 24: 474-86
19. Trimble IM, West M, Knapp M, Pownall R, Smith AFM: Detection of renal allograft rejection by computer. British medical journal 1983; 5, 286: 1695-1699.
20. Review anonymous: Kidneys and computers. British medical journal 1980: 281: 1302-3.
21. Smith AFM, West M: Monitoring renal transplants: an application of the multiprocess Kalman Filter. Biometrics 1983; 39: 867-78.
22. Kreis H: Transplanted Kidney: Natural History. In: Hamburger J, Crosnier J, Bach JF, Kreis H (eds.): Renal Transplantation. 2nd edition. Williams & Wilkins, Baltimore, 1981, pp. 177.
23. Clark EA, Terasaki PI, Opelz G, Mickey MR: Cadaver kidney transplant failures at one month. N Engl J Med. 1974; 291: 1099.
24. Skow PE, Hansen JE: The functional pattern of the cadaveric kidney in early post transplant period. Acta Med. Scand. 1974; 196:285.
25. Kjellstrand CM, Casali RE, Shideman J, Simmons RL, Buselmeir T, Najarian JS: The etiology and prognosis in acute post-transplant renal failure. Am J Med 1976, 61: 190.

26. Kreis H, Finch TW, Moreau JF, Noel LH, Lacombe M, Crosnier J: Early renal failure after cadaveric kidney transplantation. In: Hamburger J, Crosnier J, Grünfeld JP, Maxwell MH (eds.): Advances in Nephrology. Year Book Medical Publishers, Chicago, 1979, vol 8, pp 209.
27. Opelz G, Lenhard V: Immunological factors influencing renal graft survival. Annu Rev Med 1983; 34, 133-44.
28. Lenhard V, Dreikorn K, Ritz E: Neuere Aspekte der zellvermittelten und humoralen Mechanismen der Abstoßungsreaktion nach Nierentransplantation. In: Dreikorn K, Ritz E (eds.): Aktueller Stand der Nierentransplantation 1982, Nieren- und Hochdruckkrankheiten, 11, 1982; 2: 49-58.
29. Williams GM: Clinical aspects of allograft rejection. Transplant. Proc. 1974; 6:71.
30. Williams GM: Clinical course following renal transplantation. In: Morris JP (ed.): Kidney Transplantation. Principles and Practice. Academic Press, London, 1979, pp 203.
31. Anderson CB, Newton WI: Accelerated human renal allograft rejection. Arch. Surg. 1975; 110: 1230.
32. Held E, Edmaier MA: Diagnostik der Abstoßungsreaktionen. In: Albert FW, Kreiter H, Jutzler GA, Traut G (eds.): Praxis der Nierentransplantation. III. Symposium. Schattauer Verlag, Stuttgart, 1980, pp175.
33. Maury CP, Teppo AM: Comparative study of serum amyloid related protein SAA, C-reactive protein, and beta-2-microglobulin as markers of renal allograft rejection. Clin Nephrol 1984; 22(6): 284-92.
34. Woo KT, Lee EJ, Lau YK, Lim CH: Beta-2-microglobulin in the assessment of renal function of the transplanted kidney. Nephron 1985; 39 (3): 223-7.
35. Barnes RM, Alexander LC, West CR: Beta-2-microglobulin and renal graft rejection: relationship of plasma creatinine during stable transplant function and graft rejection. Transplant Proc Dec 1984; 16 (6): 1613-5.
36. Seitz R, Lutz M, Michalik R, Lange H, Klingemann HG, Egbring R: Fibronectin plasma levels after cadaver kidney transplantation. Blut 1985; 50 (1), 35-43.
37. Andrassy K, Ritz E: Stellenwert der SDS-Polyacrylamidgel-Elektrophorese bei der Diagnose von Abstoßungsreaktionen nach Nierentransplantation. In: Dreikorn K, Ritz E (eds.): Aktueller Stand der Nierentransplantation. Nieren- und Hochdruckkrankheiten, 6/1982; p 223-225.
38. Harrison JF, Barnes AD, Blainey JD: Lysozymuria and renal transplantation. Transplantation (Baltimore) Apr 1972; 13 (4): 372-7.
39. Horpacsy G et al: Changes in serum and urine lysozyme activity after kidney transplantation: Influence of graft function and therapy with azathioprine. Clinical Chemistry (Winston-Salem) 1978; 24 (1): 74-9.
40. Kutter D: Diagnostische Wertigkeit der Lysozymbestimmung im Harn. Z Med Lab Diagn 1985; 26 (1): 21-7.
41. Wellwood JM, Ellis BG, Hall JH, et al: Early warning of rejection? Br. Med. J. 1973; ii: 261
42. Ellis L, Mc Swiney RR, Tucker SM: Urinary excretion of lysozyme and N-acetyl-beta-D-Glucosaminidase in the diagnosis of renal allograft rejection. Annals of Clinical Biochemistry (Warwick-London) Sep 1978; 15 (5): 253-60.
43. Bourbouze J, Percheron F, Gluckman JC, Frantz P, Paraire M, Baumann FC, Luciani J, Legrain M: Early monitoring of human renal transplantations by NAG isoenzyme activities in urines. Clin Chim Acta Jul 1985; 149 (2-3): 185-95
44. Jung K, Diego J, Strobelt V: Diagnostic significance of urinary enzymes in detecting acute rejection crises in renal transplant recipients depending on expression of results illustrated through the example of alanine aminopeptidase. Clin Biochem Aug 1985; 18 (4): 257-60.
45. Uthmann U, Dreikorn K, Geisen HP: Der Stellenwert von Beta-2-Mikroglobulin-Bestimmungen in Serum und Urin bei der Diagnostik von Funktionsstörungen nach Nierentransplantation. In: see 37. p 84-94.
46. Fields BL, Sollinger HW, Glass NR, Carlson IH, Belzer FO: Beta-2-Microglobulin versus creatinine as the sole indicator of rejection in renal transplants. Transplantation Proceedings 1984; 6 (Dec) Vol 16, 1591-3.

47. Braun WE, Merrill JP: Urine fibrinogen fragments in human renal allografts: a possible mechanism of renal injury. N Eng. J Med 1968; 278: 1366.
48. Foegh ML, Winchester JF, Zmudha et al: Urine i-TXB$_2$ in renal allograft rejection. The Lancet 1981: 431-34.
49. Clorius JH, Dreikorn K, Horsch R, Rössler W: Die Bedeutung nuklearmedizinischer Untersuchungen mit 131-J- bzw. 123-J-Hippurat und 99mTc-Pertechnetat nach Nierentransplantation. In: see 37. 59-75.
50. Carpenter CB, Milford EL: Renal Transplantation: Immunobiology. In: Brenner BM, Rector FC (eds.): The Kidney. W.B. Saunders Company, Philadelphia, 1986, pp 1907.
51. Mihatsch MJ, Gudat F, Zollinger HU: Morphologische Kriterien der Transplantat-abstoßung. In: see 32; pp191.
52. Spichtin HP, Mihatsch MJ, Oberholzer M, Gudat F, Thiel G, Harder F, Zollinger HU: Nierenbiopsiemorphologie und Prognose des Nierentransplantates. In: see 37; 95-100.
53. Axelsen RA, Seymour AE, Methew TH, Canny A, Pascoe V: Glomerular transplant rejection: a distinctive pattern of early graft damage. Clin Nephrol Jan 1985, 23 (1): 1-11.
54. Silverman BW: Density estimation for statistics and data analysis. Chapman and Hall, London, 1986.
55. Rosenblatt,M: Remarks on some nonparametric estimates of a density function. Ann Math Statist 1956; 27: 832-837.
56. Parzen E: On estimation of a probability density and mode. Ann Math Statist 1962; 33: 1065-76.
57. Rudemo M: Empirical choice of histograms and Kernel density estimators. Scand J Statist 1982; 9: 65-78.
58. Müller HG: Boundary effects in nonparametric curve estimation models. In: Havranek T, Sidak Z, Novak M (eds.): Compstat 1984 Physica Verlag, Wien.
59. Gasser Th, Müller HG: Estimating regression functions and their derivatives by the kernel method. Scand J Statist 1984; 11: 171-84.
60. Wahba G: Smoothing noisy data with spline functions. Numer Math 1975; 24: 383-93.
61. Newsletter 27 Eurotransplant Foundation Leiden 4/85.

6. APPENDIX

Computer Programs.
This section contains a printout of the following FORTRAN subroutines
ready for implementation of the algorithm: PROGNO, PWERT, KERN4,
KEFFBO, KEWEIB, KERSOL. For documentation of the programs see comments.

```
      SUBROUTINE PROGNO( ITAG, IVOR, IVOR1, TAGE, KREA, HARN, UVOL,
     *                IKREA, IHARN, IUVOL, BKREA, BHARN, BUVOL,
     *                BIKREA, BIHARN, BIUVOL, TF, S, W, PP)
C*****************************************************************************
C    PURPOSE : EVALUATION OF ONLINE DIAGNOSIS FOR POSTTRANSPLANT
C              RENAL MONITORING.
C
C    PROGRAMMER : LOHRENGEL  -  10.2.1986
C
C    INPUT PARAMETERS
C    ==================
C         ITAG  : THE DAY TO BE ESTIMATED
C         IVOR  : NUMBER OF DAYS BEFORE THE DAY TO BE ESTIMATED
C         IVOR1 = IVOR+1
C         TAGE(IVOR) : VALUES OF THE DAYS BEFORE THE DAY TO BE ESTIM.
C         KREA(IVOR) : VALUES OF CREATININE
C         HARN(IVOR) : VALUES OF UREA
C         UVOL(IVOR) : VALUES OF URINE VOLUME
C         IKREA(IVOR) : VALUES OF INVERSE OBSERVED CREATININE
C         IHARN(IVOR) : VALUES OF INVERSE OBSERVED UREA
C         IUVOL(IVOR) : VALUES OF INVERSE OBSERVED URINE VOLUME
C         BKREA  : BANDWIDTH FOR CREATININE
C         BHARN  : BANDWIDTH FOR UREA
C         BUVOL  : BANDWIDTH FOR URINE VOLUME
C         BIKREA : BANDWIDTH FOR INVERSE OBSERVED CREATININE
C         BIHARN : BANDWIDTH FOR INVERSE OBSERVED UREA
C         BIUVOL : BANDWIDTH FOR INVERSE OBSERVED URINE VOLUME
C         TF     : TUNING FACTOR TO MODIFY BANDWIDTHS
C    OUTPUT-PARAMETER
C    =================
C         PP     : P - VALUE
C
C    SCRATCH PARAMETERS
C    ==================
C         S(IVOR1),W(IVOR) : HAVE TO BE CREATED IN MAIN PROGRAM
C         XOUT(1) , YOUT(1) , VAR(1) , PARRAY(8)
C         IKR30  : INVERSE OBSERVED CREATININE AT CURRENT DAY
C         IUV3A2 : KERNEL ESTIMATED SECOND DERIVATIVE OF
C                  INVERSE URINE VOLUME AT CURRENT DAY
C         HST3A1 : KERNEL ESTIMATED FIRST DERIVATIVE OF UREA
C                  AT CURRENT DAY
C         IHST2D : DIFFERENCE OF KERNEL ESTIMATED AND OBSERVED INVERSE
C                  UREA AT (CURRENT DAY - 1)
C         IKR230 : PRODUCT OF INVERSE OBSERVED CREATININE AT CURRENT
C                  DAY AND (CURRENT DAY - 1).
C         IKROD  : DIFFERENCE OF INVERSE OBSERVED CREATININE AT CURR.
C                  DAY MINUS (CURRENT DAY - 1)
C         UV3A   : PRODUCT OF KERNEL ESTIMATED FIRST DERIVATIVE AND
C                  SECOND DERIVATIVE OF URINE VOLUME AT CURRENT DAY
C         IHK3A1 : PRODUCT OF KERNEL ESTIMATED FIRST DERIVATIVES OF
C                  INVERSE UREA AND INVERSE CREATININE AT CURRENT DAY
C
C     SUBROUTINES USED : PWERT , KERN4  KERSOL , KEFFBO , KEWEIB
```

```
      DIMENSION HARN(IVOR),UVOL(IVOR),XOUT(1),YOUT(1),S(IVOR1)
      DIMENSION W(IVOR),VAR(1),PARRAY(8),TAGE(IVOR)
      REAL IKREA(IVOR),IHARN(IVOR),IUVOL(IVOR),KREA(IVOR)
      REAL IKR30,IUV3A2,IHST2D,IKR230,IKROD,IHK3A1
      IKR30 = IKREA(IVOR)
C
C     COMPUTATION OF IUV3A2
C
      NUE = 2
      WID = TF * BIUVOL
      XOUT(1) = FLOAT(ITAG)

      CALL KERN4(NUE, WID,IVOR,IVOR1,TAGE,IUVOL,S,W,
     *           XOUT, YOUT, VAR)
      IUV3A2 = YOUT(1)
C
C     COMPUTATION OF HST3A1
C
      NUE = 1
      WID = TF * BHARN
      XOUT(1) = FLOAT(ITAG)
      CALL KERN4(NUE, WID,IVOR,IVOR1,TAGE,HARN,S,W,
     *           XOUT, YOUT, VAR)
      HST3A1 = YOUT(1)
C
C     COMPUTATION OF IHST2D
C
      NUE = 0
      WID = TF * BIHARN
      XOUT(1) = FLOAT(ITAG - 1)
      CALL KERN4(NUE, WID,IVOR,IVOR1,TAGE,IHARN,S,W,
     *           XOUT, YOUT, VAR)
      IHST2D = IHARN(IVOR-1) - YOUT(1)
      IKR230 = IKREA(IVOR-1) * IKREA(IVOR)
      IKROD = IKREA(IVOR) - IKREA(IVOR-1)
C
C     COMPUTATION OF UV3A
C
      NUE = 1
      WID = TF * BUVOL
      XOUT(1) = FLOAT(ITAG)
      CALL KERN4(NUE, WID,IVOR,IVOR1,TAGE,UVOL,S,W,
     *           XOUT, YOUT, VAR)
      Y = YOUT(1)
      NUE = 2
      WID = TF * BUVOL
      XOUT(1) = FLOAT(ITAG)
      CALL KERN4(NUE, WID,IVOR,IVOR1,TAGE,UVOL,S,W,
     *           XOUT, YOUT, VAR)
      UV3A = Y * YOUT(1)
C
C     COMPUTATION OF IHK3A1
C
      NUE = 1
      WID = TF * BIKREA
      XOUT(1) = FLOAT(ITAG)
      CALL KERN4(NUE, WID,IVOR,IVOR1,TAGE,IKREA,S,W,
     *           XOUT, YOUT, VAR)
      Y = YOUT(1)
      WID = TF * BIHARN
      CALL KERN4(NUE, WID,IVOR,IVOR1,TAGE,IHARN,S,W,
     *           XOUT, YOUT, VAR)
      IHK3A1 = Y * YOUT(1)
```

```
      PARRAY(1) = IKR30
      PARRAY(2) = IUV3A2
      PARRAY(3) = HST3A1
      PARRAY(4) = IHST2D
      PARRAY(5) = IKR230
      PARRAY(6) = IKROD
      PARRAY(7) = UV3A
      PARRAY(8) = IHK3A1
      IDIM = 8
C
C    CALCULATION OF P-VALUE
C
      CALL PWERT( PARRAY, IDIM, PP)
      RETURN
      END
C********************************************************************
      SUBROUTINE PWERT( X, N, PP)
C********************************************************************
C   EVALUATES LOGISTIC FUNCTION TO OBTAIN P-VALUE
C
C   INPUT-PARAMETERS
C   ================
C
C   X(N)  :  THE VALUES OF IKR30 , IUV3A2 , HST3A1 , IHST2D , IKR230 ,
C            IKROD , UV3A , IHK3A1
C
C   OUTPUT-PARAMETER  :
C   ===================
C   PP   :   P - VALUE
C
C   SCRATCH-PARAMETER
C   =================
C
C   COEFF(9) : COEFFICIENTS OF LOGISTIC MODEL
C
      DIMENSION X(N),COEFF(9)
      DATA COEFF / -54.7, -11000., -0.0853, 3520., 89.0, 143.0,
     *             -0.00000704, -46400.0, 11.7 /
      U = COEFF(9)
      DO 10 I=1,N
      U = U + COEFF(I) * X(I)
   10 CONTINUE
      PP = 1.0
      IF(U.LT.20.0) PP = EXP(U) / (1.0 + EXP(U))
      RETURN
      END
C********************************************************************
      SUBROUTINE KERN4(NUE,WID,N,N1,T,X,S,W,U,Y,VAR)
C********************************************************************
C
C      *** PROGRAMMER : CH. LOHRENGEL / H.G. MUELLER ***   5.12.1985
C
C      PURPOSE: KERNEL SMOOTHING ROUTINE FOR PROGNOSIS AT ONE POINT,
C               INCLUDING DERIVATES   NUE = 0,1,2 .
C      METHOD:  USE OF BOUNDARY KERNELS
C
C      REMARKS:  KERNELS USED (NKE = 2) HAVE ORDER   KORD = NUE + 2
C
C      PARAMETERS:
C      ----------
C      INPUT NUE    ORDER OF DERIVATIVE TO BE ESTIMATED (0,1 OR 2).
C      INPUT N      NUMBER OF DATA (N>3 REQUIRED)
```

203

```
C          INPUT N1      N1=N+1
C          INPUT T(N)    POINTS WHERE DATA HAVE BEEN SAMPLED (MUST BE MONOT.
C          INPUT X(N)    VECTOR OF DATA
C          INPUT U(1)    POINT WHERE CURVE IS TO BE ESTIMATED
C                        IT IS REQUIRED THAT U(1) GE T(N)
C          INPUT  WID    BANDWIDTH TO BE USED
C          OUTPUT Y(1) ESTIMATE AT U(1)
C          OUTPUT VAR(1) COMPUTED VARIANCE (UP TO A FACTOR SIGMA**2) AT U1
C          SCRATCH S(N1)  VECTOR OF INTERPOLATING SEQUENCE (N1=N+1)
C          SCRATCH W(N)  VECTOR OF WEIGHTS
C
C          ARRAYS NEEDED IN MAIN PROGRAM:
C          ------------------------------
C          T(N),X(N),W(N),S(N1),U(1),Y(1),VAR(1)
C          DOUBLE PRECISION C(20)
C
C          SUBROUTINES USED: KEWEIB,KERSOL,KEFFBO
C
           DIMENSION X(N),T(N),U(1),Y(1),S(N1),W(N),VAR(1)
           DOUBLE PRECISION WNUE
           DOUBLE PRECISION C(20)
C
           NKE = 2
           KORD = NUE + 2
           WNUE=DBLE(1.0)
           IF(NUE.GT.0) WNUE = DBLE(WID**NUE)
           LOW=1
           NK = NKE - 1
           IOR = KORD + 1 + 2*NK
           NB = -1
           J = 1
           SO=0.0
           UU=U(J)
           UUN=UU-WID
C
C      COMPUTATION OF RELEVANT INDICES
C
   120 IF(S(LOW).GT.UUN) GOTO 140
           LOW=LOW+1
           GOTO 120
C 140 IF(S(LOW).GT.UU) CALL KERERR(8)
   140 LHI=LOW-1
   160 LHI=LHI+1
           IF(LHI.EQ.N1) GOTO 180
           IF(S(LHI).LT.UU) GOTO 160
   180 IF(LOW.GT.1) LOW=LOW-1
           IF(LHI.GT.2) LHI = LHI - 1
C          IF(LHI.LE.2.OR.LOW+1.GT.LHI) CALL KERERR(12)
           SX = (S(LHI)-T(LHI-1)) / 2.0
           SS = S(LHI)
           S(LHI) = S(LHI) - SX
           Q = (S(LHI) - UU) / WID
           LHI = LHI - 1
           IZ = LHI - LOW + 1
C
C      COMPUTATION OF SMOOTHING POLYNOMIAL
C
           CALL KEFFBO(NKE,Q,NUE,KORD,NB,C)
           DO 60 I1=1,IOR
           C(I1) = C(I1) / WNUE
    60 CONTINUE
```

```
C
C     COMPUTATION OF WEIGHTS
C
      CALL KEWEIB(S,N1,UU,LOW,IZ,WID,C,IOR,N,W,Q,NB)
C
C     COMPUTATION OF SMOOTHED VALUE Y(1)
C
      S(LHI+1) = SS
      SUM=0.
      IZ=1
      DO 410 I1=LOW,LHI
      SUM=SUM+X(I1)*W(IZ)
      SO=SO+W(IZ)*W(IZ)
      IZ=IZ+1
  410 CONTINUE
      Y(J)=SUM
      VAR(J)=SO
 1000 CONTINUE
      RETURN
      END.
C*******************************************************************************
      SUBROUTINE KEWEIB(S,N1,UU,LOW,IZ,WID,C1,IORD,N,W,Q,NB)
C*******************************************************************************
C     VERSION FEB. 1983   ** H.G.MUELLER **
C     COMPUTED WEIGHTS FOR KERNEL ESTIMATES OF REGRESSION AND OF
C     ITS DERIVATES .
C
C     PARAMETERS
C     ----------
C     INPUT     S(N1)   INTERPOLATION SEQUENCE
C     INPUT     N1      LENGTH OF S (= N+1 , N = NUMBER OF DATA POINTS)
C     INPUT     UU      ABSCISSA TO BE ESTIMATED
C     INPUT     LOW     FIRST NON-ZERO WEIGHT
C     INPUT     IZ      NUMBER OF NON-ZERO WEIGHTS
C     INPUT     WID     BANDWIDTH
C     INPUT     C1(20)  COEFFICIENTS OF KERNEL POLYNOMIAL
C     INPUT     IORD    ORDER OF KERNEL POLYNOMIAL
C     INPUT     N       NUMBER OF DATA POINTS
C     OUTPUT    W(N)    WEIGHTS FOR THE ESTIMATION AT UU
C
      DIMENSION S(N1),W(N)
      DOUBLE PRECISION C1(20)
C
C
      DO 1000 I1=1,IZ
         II=LOW-1+I1
         XX=(UU-S(II))/WID
         YY=(UU-S(II+1))/WID
         IF(XX.GT.1.)XX=1.
         IF(YY.LT.-Q) YY=-Q
         XXX=XX
         YYY=YY
         SUM=0.
         DO 100 I2=1,IORD
            SUM=SUM+C1(I2)*(XXX-YYY)
            XXX=XXX*XX
            YYY=YYY*YY
         IF(ABS(XXX).LT.1.0E-37) XXX=0.
         IF(ABS(YYY).LT.1.0E-37) YYY=0.
  100    CONTINUE
         W(I1)=SUM
 1000    CONTINUE
```

205

```
      RETURN
      END
C**********************************************************************
      SUBROUTINE KEFFBO(NKE,Q,NUE,KORD,NB,C1)
C**********************************************************************
C     PURPOSE: COMPUTES COEFFICIENTS OF KERNEL POLYNOMIAL
C     -------        AT BOUNDARY
C     PARAMETERS:  INPUT SEE KESMO
C     ----------       OUTPUT C1(7)  COEFFICIENTS OF POLYNOMIAL
      DOUBLE PRECISION A(7,7),SL(7),R(7)
            DOUBLE PRECISION C1(20)
      COMMON A
      DO 5 I=1,20
      C1(I)=0.
5     CONTINUE
C
C     CONSTRUCTION OF A
C
      KA=KORD
      KK=KORD
      IF(NKE.EQ.2) KA=KORD+1
      A(1,1)=DBLE(Q)
      DO 10 I=2,KA
      A(1,I)=DBLE(Q)*A(1,I-1)
10    CONTINUE
      DO 20 I=2,KK
      A(I,KA)=A(I-1,KA)*DBLE(Q)
      IC=I+KA-1
      IF(MOD(IC,2).EQ.0) A(I-1,KA)=(1.0+A(I-1,KA))/FLOAT(IC-1)
      IF(MOD(IC,2).NE.0) A(I-1,KA)=(1.0-A(I-1,KA))/FLOAT(IC-1)
      IF(MOD(IC,2).NE.0.AND.NB.GT.0) A(I-1,KA)=-A(I-1,KA)
20    CONTINUE
      IC=KK+KA
      IF(MOD(IC,2).EQ.0) A(KK,KA)=(1.0+A(KK,KA))/FLOAT(IC-1)
      IF(MOD(IC,2).NE.0) A(KK,KA)=(1.0-A(KK,KA))/FLOAT(IC-1)
      IF(MOD(IC,2).NE.0.AND.NB.GT.0) A(KK,KA)=-A(KK,KA)
      KM=KA-1
      DO 30 I=1,KM
      IF(MOD(I+1,2).EQ.0) A(1,I)=(1.0+A(1,I))/FLOAT(I)
      IF(MOD(I+1,2).NE.0) A(1,I)=(1.0-A(1,I))/FLOAT(I)
      IF(MOD(I+1,2).NE.0.AND.NB.GT.0) A(1,I)=-A(1,I)
30    CONTINUE
      DO 40 I=2,KK
      DO 50 J=1,KM
      LC=I+J-1
      LCC=I+J-KA
      IF(LC.LE.KA) A(I,J)=A(1,LC)
      IF(LC.GT.KA) A(I,J)=A(LCC,KA)
50    CONTINUE
40    CONTINUE
      IF(NKE.EQ.1) GOTO 100
      DO 60 I=1,KK
      DO 70 J=1,KA
      IN=KA-I+1
      INN=IN-1
      A(IN,J)=A(INN,J)
70    CONTINUE
60    CONTINUE
      DO 80 I=1,KA
      A(1,I)=DBLE(1.0)
      IF(NB.GT.0.AND.MOD(I,2).EQ.0) A(1,I)=-A(1,I)
80    CONTINUE
```

```
100     NUU=NUE+1
        IF(NKE.EQ.2)NUU=NUE+2
        IF(NUE.EQ.0) F=1.
        IF(NUE.EQ.1) F=-1.
        IF(NUE.EQ.2) F=2.
        DO 110 I=1,KA
        R(I)=DBLE(0.)
110     CONTINUE
        R(NUU)=DBLE(F)
        CALL KERSOL(KA,R,SL)
        DO 120 I=1,KA
        C1(I)=SL(I)/FLOAT(I)
120     CONTINUE
        RETURN
        END
C*******************************************************************
        SUBROUTINE KERSOL(ND,R,SL)
C*******************************************************************
C       SUBROUTINE FOR SOLUTION OF LINEAR SYSTEM
C       A*SL=R   WITH PIVOTING
C       OF DIMENSION ND.
C       PROGRAMMER : H.G.MUELLER , 1982
C       ARRAYS NEEDED:
C       NL(7)                 MEMORY ARRAY FOR PERMUTATION OF COLUMNS
C       DOUBLE PRECISION A(7,7),SL(7),R(7)    PARTS OF LINEAR SYSTEM
C
C       SCRATCH NF   NF=1 IF SYSTEM IS DEGENERATE (SHOULD NOT
C                    OCCUR IF KERSOL IS CALLED BY KEFFBO)
        DIMENSION NL(7)
        DOUBLE PRECISION A(7,7),R(7),SL(7)
        DOUBLE PRECISION SUM,MAX,AA,MS,RE
        COMMON A
        NF=0
        DO 5   J =1,ND
        NL(J) =J
5       CONTINUE
C
C       SEARCH OF MAXIMAL PIVOT ELEMENT
C
        NDD= ND-1
        DO 10 J1=1,NDD
        MAX=DBLE(0.0)
        KS=J1+1
        K0=J1
        L0=J1
        DO 20 J2=J1,ND
        SUM=DBLE(0.0)
        DO 30 J3=1,ND
        SUM=SUM+DABS(A(J2,J3))
30       CONTINUE
        IF (SUM.EQ.0.0) GOTO 20
        DO 40 J4=J1,ND
        AA=A(J2,J4)
        IF(AA.LT.0.0)  AA=-AA
        MS=MAX*SUM
        IF(AA.LE.MS) GOTO 40
        MAX=AA/SUM
        K0=J2
        L0=J4
40      CONTINUE
20      CONTINUE
```

```
        IF (MAX.EQ.0.0) GOTO 500
C
C       PERMUTATION OF LINE AND COLUMN
C
        IF(K0.EQ.J1) GOTO 60
        RE=R(K0)
        R(K0)=R(J1)
        R(J1)=RE
        DO 50 J5=1,ND
        AA=A(J1,J5)
        A(J1,J5)=A(K0,J5)
        A(K0,J5)=AA
50      CONTINUE
60      IF (L0.EQ.J1) GOTO 80
        ID=NL(L0)
        NL(L0)=NL(J1)
        NL(J1)=ID
        DO 70 J7=1,ND
        AA=A(J7,J1)
        A(J7,J1)=A(J7,L0)
        A(J7,L0)=AA
70      CONTINUE
C
C       MODIFICATION OF MATRIX A
C
        IF (A(J1,J1).EQ.0.0) GOTO 500
80      DO 90 J9=KS,ND
        SL(J9)=-A(J9,J1)/A(J1,J1)
        R(J9)=R(J9)+SL(J9)*R(J1)
        DO 100 J10=KS,ND
        A(J9,J10)=A(J9,J10)+SL(J9)*A(J1,J10)
100     CONTINUE
90      CONTINUE
10      CONTINUE
C
C       COMPUTATION OF SOLUTION SL
C
200     DO 210 J=1,ND
        SUM=0.0
        J1=ND-J+1
        KJ=J1+1
        IF(J1.LT.ND) GOTO 220
        IF (A(J1,J1).EQ.0.0) GOTO 500
        SL(J1)=R(J1)/A(J1,J1)
        GOTO 210
220     DO 230 J2=KJ,ND
        SUM=SUM+A(J1,J2)*SL(J2)
230     CONTINUE
        IF (A(J1,J1).EQ.0.0) GOTO 500
        SL(J1)=(R(J1)-SUM)/A(J1,J1)
210     CONTINUE
        DO 300 J=1,ND
        ID=NL(J)
        R(ID)=SL(J)
300     CONTINUE
        DO 310 J=1,ND
        SL(J)=R(J)
310     CONTINUE
        RETURN
500     NF=1
        RETURN
        END
```

ASSISTING THE EXPERT IN "EXPERT" SYSTEMS

Herman P. Friedman
IBM Corporate Technical Institutes
Systems Research Institute
500 Columbus Avenue
Thronwood, N.Y. 10594
USA

ABSTRACT

Expert Systems are at the cutting edge of Artificial Intelligence. A number of successes have been claimed in problem areas usually associated with human decision making.

The purpose of this paper is to present some concepts and tools that have relevance to the construction and evaluation of "Expert" systems. Results of studies in multivariate data analysis, as well as studies of biases in human judgment in context where the outcome is uncertain (i.e., classification medicine) will be presented. Implications for the development of Expert Systems, as Decision Support Systems, will be drawn from these studies.

INTRODUCTION

We distinguish between two goals of Artificial Intelligence. One goal is to develop computer models in order to better understand human cognition and decision making. This activity is basic research and the models attempt to describe or explain how people make decisions. The other goal is more technology oriented to solve problems. It is the point of view of this paper that the goal of "Expert Systems" is to solve problems. In the context of this paper the problems involve decision making in the face of uncertainty in the domains of medicine involving diagnosis, treatment and prognosis. A frame of reference needs to be established before the decision problem can be adequately formulated. Some examples of the collaborative roles of physicians and statistical scientists in the development of such frames of reference are provided by H. P. Friedman (1987).

In this sequel we examine the consequences of a rule based approach to building "Expert Systems" and conclude that it is better to use the computer to assist the Expert in his job rather than to do the job completely and inadequately. It is easier and better for a computer to examine the reasonableness and correctness of a diagnosis or treatment plans created by a physician than to create them in the first place.

DISCUSSION AND RECOMMENDATIONS

Many studies have shown that human beings are not very good at probability and risk assessment. See for discussion, (D. Kahneman and A. Tversky (1982)).

In addition human reasoning cannot adequately be described in terms of content independent of formal rules. The actual reasoning process is schema-bound or context bound. Different operations or inferential rules are available in different contexts: G. Shafer (1985) has indicated that people do not come to the task of probability judgment with well-structured beliefs hidden in their psyches waiting to be elicited. Probabilities need to be constructed rather than elicited. Contrast this idea with the goal of expert systems design to "elicit" rules from the expert and program them into the computer.

In the book by D. Kahneman and A. Tversky (1982), there are a number of papers that further enumerate and describe heuristics and biases associated with human judgment under uncertainty. In particular they are: representativeness, availability, insensitivity to prior probability, incoherence, insensitivity to sample size, misconceptions of regression, as well as misuse of causality and attribution.

When we are concerned with technological aids for decision making should we be modelling and reproducing these human failings?

There is a critically important intuitive component provided by the expert physician in most decisions. We do not want to lose this. However, we think the computer can be used to assist by helping to calibrate the judgment of the expert. A case study of such an approach complete with an evaluation of performance is presented by Goldman, L., et al (1982). This study developed a protocol to assist the physician in the diagnosis of emergency room patients with acute chest pain. It was developed outside the traditional context of expert systems.

There is some controversy as to the appropriate normative model for judgment under uncertainty. The major choices are Bayesian Probability, Belief Functions and Fuzzy Logic. The interested reader will find the issues discussed in Kanal, L., and Lemmer (1986) and Henrion (1986). My own view is that all of these approaches will have to be evaluated in terms of an empirical theory of probability. This is, evaluation will be based on factual statements about performance in the real world.

Within the context of expert systems I think an approach for the near future may be found in the prize winning paper by Cooper, G. F. (1986). Since this paper was presented, a paper discussing Artificial Intelligence in medicine appeared in the New England Journal of Medicine, Szolovits, P. (1987).

REFERENCES

Cooper, G. F. (1986), "A Diagnostic Method that Uses Casual Knowledge and Linear Programming in the Application of Bayes' Formula." Computer Methods and Programs in Biomedicine 22 (1986) 223-237.

Friedman, H. P. (1987), "Strategies for Multivariate Data Analysis: Case Studies." Proceedings of International Symposium on "Acquisition Analysis and Use of Clinical Transplant Data." Heidelberg Published by Springer-Verlag (1987).

Henrion, M. (1986), "Should We Use Probability in Uncertain Inference Systems?" Proceedings of the Eighth Annual Conference of the Cognitive Science Society, Amherst, Mass., August, 1986. published by Lawrence Erblaum Associates.

Goldman, et al. (1982), "A Computer-Derived Protocol to Aid in the Diagnosis of Emergency Room Patients With Acute Chest Pain," The New England Journal of Medicine, Vol. 307, No. 10, pp. 588-596, September 2, 1982.

Kahneman, D. and Tversky, A. (Editors with P. Slovic), Judgment under uncertainty: Heuristics and biases, Cambridge University Press (1982).

Kanal, L. M., Lemmer, J. F. (1986) (Editors), Uncertainty in Artificial Intelligence, North-Holland (1986).

Szolovits, P. (1987), "Artificial Intelligence in Medicine: Where do we stand.", New England Journal of Medicine, March 12, 1987, Vol. 316, No. 11, pp. 685-688.

CLINICAL MARROW TRANSPLANTATION

John A. Hansen, Patrick G. Beatty, Paul J. Martin
and E. Donnall Thomas
The Histocompatibility Laboratory, Puget Sound Blood Center; The
Division of Clinical Research, Fred Hutchinson Cancer Research Center;
and The Department of Medicine, Division of Oncology, University of
Washington School of Medicine
Seattle, WA

I.   INTRODUCTION

Transplantation of normal blood-forming marrow stem cells from
one individual to another, excluding an identical twin, is referred to
as an allogeneic transplant (1,2). When the grafted donor cells and
the recipient are foreign to each other, certain immunological
reactions can occur and cause clinically significant complications
(3,4). Marrow transplanted between identical twins causes no
significant reactions (5).   Graft-versus-host disease represents a
complication unique to allogeneic marrow grafts, but graft rejection
can also occur. Rehabilitation of the marrow transplant patient
requires full reconstitution of the hematopoietic system (bone

---

Address correspondence to John A. Hansen, M.D., Fred Hutchinson Cancer
Research Center, 1124 Columbia Street, Seattle, WA 98104, 296/292-
6545. This work was supported by National Institutes of Health grants
CA 15704 and CA 18029 from the National Cancer Institute and HL 36444
(formerly CA 30924) and HL 33775 from the National Heart, Lung and
Blood Institute. E.D. Thomas is the recipient of Research Career
Award AI 02425 from the National Institute of Allergy and Infectious
Diseases.

marrow and lymphoid), establishment of donor-recipient tolerance, and normalization of the immune system and resistance to infection.

Indications. An allogeneic marrow transplant is the treatment of choice for several congenital and acquired hematologic diseases (Table 1). Transplants from a normal HLA genotypically identical sibling or HLA phenotypically identical parent are lifesaving for children with severe combined immunodeficiency disease (SCID) (6,7). Lacking an HLA identical donor, children with SCID have been successfully treated by transplanting T cell-depleted marrow from an HLA incompatible parent (8). Patients with severe aplastic anemia (SAA), a disease of the bone marrow resulting in failure of formation of new red cells, white cells and platelets, are at high risk of fatal infection and bleeding. Supportive care alone is successful in sustaining life in < 30% SAA patients, but treatment with an HLA identical marrow transplant can be curative (9-11). Long-term survival is > 80% when SAA patients are transplanted early in the course of their disease before transfusion-induced alloimmunization has occurred (Table 2). Two genetic diseases associated with impaired hematopoiesis and bone marrow failure, thalassemia and Fanconi's anemia, also have been corrected by marrow transplants (13,14).

TABLE 1. CLINICAL INDICATIONS FOR MARROW TRANSPLANT

I. Established Indications (HLA Identical Transplants)

    o Severe combined immunodeficiency disease (SCID)
    o Severe aplastic anemia (SAA)
    o Thalassemia
    o Fanconi's anemia
    o Chronic myelogenous leukemia (CML)
    o Acute lymphocytic leukemia (ALL)
        high risk 1st CR, relapse, 2nd CR
    o Acute nonlymphocytic leukemia (ANL)
        1st CR, relapse

II. Active Clinical Trials

    o HLA partially matched donors
    o Unrelated donors
    o Lymphoma
    o Inborn errors of metabolism:
        Gaucher's disease, Hurler's syndrome, etc.
    o AIDS

III. Potential Applications

    o Hemoglobinopathies
    o Autoimmune disorders
    o Radiation injury

Allogeneic marrow transplants offer the potential for cure for certain patients with leukemia (15-18). The optimal time for undertaking a transplant varies for different types of leukemia depending on the expected rate of progression, risk of relapse and effectiveness of conventional chemotherapy. Since progress will occur in conventional chemotherapy as well as marrow transplantation, assessment of the relative advantages of the two approaches requires randomized prospective trials (19,20). Clinical trials are also underway to determine the efficacy of marrow transplants for treating other malignant diseases such as malignant lymphoma (21).

Marrow transplantation combined with anti-viral agents may be effective treatment for AIDS. With continued improvement in marrow transplant procedures, it should be feasible to consider using normal hematopoietic stem cells for treatment of several other diseases such as hemaglobinopathies, inborn errors of metabolism, and certain autoimmune disorders. Finally, marrow transplants might be effective treatment for rescuing individuals exposed to life-threatening doses of radiation. Absence of significant thermal and traumatic injuries, limited radiation injury to other tissues, and speed in finding a matched donor will be relevant factors in determining whether allogeneic marrow transplants can be useful in this setting (22).

TABLE 2. SUMMARY OF THE LONG-TERM EXPERIENCE OF THE SEATTLE TEAM WITH A VARIETY OF PATIENTS GIVEN A MARROW TRANSPLANT FROM AN HLA IDENTICAL SIBLING DONOR

Disease Category	No. of Patients	Long-Term Disease-Free Survivors	Actuarial Long-Term Disease-Free Survivors	Time from Transplant (Years)
Aplastic anemia:				
Transfused pre-1976	80	37	46%[*]	9-14
Untransfused	39	32	82%	5-12
Transfused 1976-81	42	28	67%	5-9
Fanconi's anemia	10	5	-	2-11
Thalassema major	7	5	-	1-5
End-stage acute leukemia	100	11	11%	10-14
ALL in 2nd remission	46	14	26%	5-9
ANL in 1st remission	107	47	48%	5-9
CML, blastic phase	42	4	12%	2-8
" accelerated phase	46	11	16%	2-5
" chronic phase	67	36	50%	2-7
Adv. Hodgkin's & lymphoma	100	30	18%	2-14

[*] The per cent of patients that appear to be on a long-term plateau when analyzed by the method of Kaplan and Meier (see ref. 12)

II. TRANSPLANT-RELATED COMPLICATIONS

Graft Rejection. Stimulation of the recipient by alloantigens expressed on cells of the donor marrow can lead to graft rejection; however, this problem can be overcome by immunosuppression (1). In certain high-risk groups such as transfused SAA patients, additional immunosuppression is needed (9). Patients transplanted for hematologic malignancy usually receive chemotherapy and total body irradiation (TBI) at high doses to achieve the maximum tolerable anti-tumor effect (2). Such therapy is marrow-albative and usually sufficiently immunosuppressive to prevent graft rejection. Nevertheless, graft rejection still occurs in some leukemic patients transplanted from HLA incompatible donors (23).

Graft-vs-Host Disease (GVHD). The ability of donor T cells in the transplanted marrow to recognize foreign antigens of the recipient is the basis for a graft-versus-host reaction (1,4,5). If sufficiently strong, this reaction can cause a constellation of problems known as graft-versus-host disease (GVHD). Since significant GVHD does not occur in identical twin transplants, the genetic differences responsible are known as histocompatibility (H) antigens. The strongest or major-H antigens are linked to the HLA system, disparity for HLA can cause severe GVHD (4,23). Clinically significant GVHD, however, also occurs in some HLA identical transplants indicating that certain non-HLA minor-H antigens may also be pathogenic.

Depending on relevant variables such as patient age and the type of GVHD prophylaxis employed, the incidence of clinically significant GVHD in HLA identical transplants is approximately 30-50% (1-3,24-27). The risk of death from GVHD and related complications is approximately 5-20%. Depletion in vitro of T cells from donor marrow pretransplant represents an important alternative approach for preventing GVHD (8,28-31). T cell depletion offers the advantage of not requiring immunosuppressive drugs after transplantation, but unfortunately has been associated with an increased risk of graft rejection (28,30,31).

Immune Deficiency. Complete recovery of cell-mediated and
humoral immunity after marrow transplantation can be delayed for
several months. During this time there is an increased susceptibility
to certain infections, particularly those caused by herpes simplex,
cytomegalovirus, pneumocystis carinii and varicella zoster (32).
Persistent GVHD and infection can themselves be immunosuppressive.
Long-term followup studies indicate that tests for monitoring immune
function do not become completely normal until at least 12 months
after transplantation, even in clinically well patients (33).

III. CRITERIA FOR DONOR SELECTION

The absence of a normal identical twin or HLA genotypically
identical sibling does not preclude marrow transplantation. The
alternatives are an autologous transplant (if feasible), a transplant
from an HLA partially matched family member (34), or a transplant from
an HLA identical unrelated volunteer (35,36).

HLA partially matched family members can be identified when there
is fortuitous sharing of parental HLA antigens. This can occur by
chance or by inbreeding. Two examples are illustrated in Figure 1.
In family A, the parents share similar haplotypes and the first sib is
HLA matched with his father for HLA-B and -D but mismatched for the A
locus (haplotypes "b" and "c"). Since sib #1 and his father share a
common haplotype (the paternal "a" haplotype), this match is
classified as a haploidentical, parent-child, one locus incompatible
transplant. A similar relationship is illustrated between the mother
and sib #2. In family B, the mother is homozygous for HLA-B8,Dw3.
Sib #1 and Sib #2 share the same paternal haplotype ("a") and although
they have inherited different maternal haplotypes ("c" and "d"), they
are only mismatched for the A locus (Al vs A31). This match is
classified as a haploidentical, sib-sib, one locus incompatible
transplant.

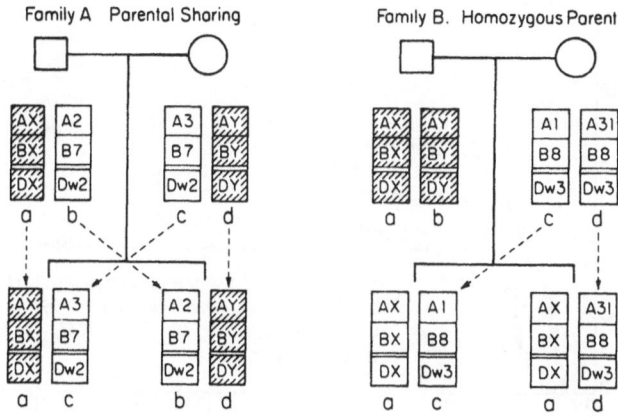

FIGURE 1.   SHARING OF HLA ANTIGENS WITHIN A FAMILY

IV. SIGNIFICANCE OF HLA COMPATIBILITY

    Transplants from Related Donors. Incompatibility for HLA
increases the risk of GVHD (23).   Data illustrated in Figure 2
represent marrow transplants in patients with hematologic malignancy.
Transplantation from a one locus (HLA-A,B or D) incompatible donor is
associated with an earlier onset and increased overall incidence of
GVHD compared to transplants from an HLA genotypically identical
donor. Patients given marrow from a haploidentical zero locus (HLA-
A,B and D) incompatible donor do not have significantly increased GVHD
compared to controls. These data indicate that phenotypic
matching for the HLA-A,B and D antigens does not result in a
detectable increase in GVHD risk. In contrast, mismatching for a
single HLA-A,B or D locus antigen provides sufficient genetic
disparity to cause a more vigorous and more frequent GVHD.   In
addition to increasing the risk of GVHD, HLA disparity also increases
the risk of graft failure and rejection (23).

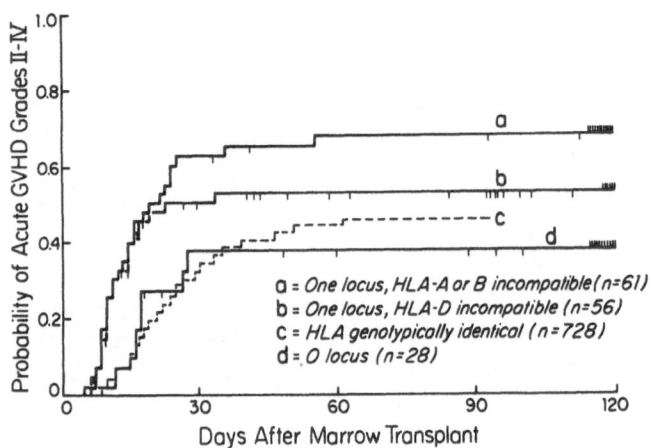

FIGURE 2.   INCIDENCE OF CLINICALLY SIGNIFICANT ACUTE GVHD IN
HLA-INCOMPATIBLE MARROW TRANSPLANTS

Transplants from Unrelated Donors.   The risk of graft rejection
and GVHD in unrelated HLA-A,B and D identical transplants is unknown
since the number of such transplants is too small for statistical
analysis.   It is theoretically possible that there will be more GVHD
in transplants from unrelated donors in spite of matching for HLA-A,B
and D because unrelated donors and recipients will share fewer non-HLA
minor-H antigens than related donors and recipients.   Furthermore,
technical errors in HLA typing may prejudice the matching of unrelated
individuals, especially when donors and patients originate from
distinct racial groups.   Constraints on the precision of HLA typing
may increase both false positive and false negative errors.

V.   PROBABILITY OF FINDING MATCHED DONORS

Each individual inherits one of two maternal HLA haplotypes and
one of two paternal HLA haplotypes.   Thus the chance that any sibling
will be HLA identical is 1 in 4.   The probability of finding an HLA
identical sibling increases in larger families.   In the U.S.A., a
patient up to the age of 18 has a 29% chance of being in a family with
at least one HLA-identical sibling (Table 3).   As the average size of

families continues to decrease, however, the fraction of patients with
an HLA identical sibling will also decrease. The probability of
finding a partially matched donor in a family (assuming that there are
two available and unrelated parents) depends primarily on the
frequency of the parental antigens. The average probability that a
child will be incompatible for only one locus (HLA-A,B or D) with
either parent is 5%, and the average probability that a child will be
incompatible for only two loci with either parent is 35%. If it were
feasible to achieve successful transplants from haploidentical two
locus incompatible donors, the problem of donor availability could be
substantially improved. Unfortunately, with current methods, the
clinical problems encountered in two locus incompatible transplants
are formidable.

TABLE 3. PROBABILITY OF FINDING HLA IDENTICAL SIBLINGS
IN U.S.A. FAMILIES

Offspring <18 Per Family	Size Distribution of Families by Size	Probability of Finding an HLA Identical Sib
0	31,594 (50%)	–
1	13,108 (21%)	–
2	11,645 (19%)	.25
3	4,486 (7%)	.44
4	1,329 (2%)	.58
5	373 (1%)	.68
≥ 6	171 (0.3%)	.76

[a] Total U.S.A. families x 10^3. Calculations based on 1985 U.S.A.
census data indicating that 29% of individuals under 18 years of
age have an HLA-identical sibling also under 18 years of age.
Data from Current Population Reports (see ref. 37).

If a patient lacks an HLA identical or partially matched related
donor, a search for an unrelated donor may be undertaken. Since 1980,
we have requested unrelated donor searches for 228 patients. Searches
within a registry of approximately 20,000 volunteers have successfully
identified at least one HLA-A,B-matched donor for 150 patients (66%).
When HLA-DR typing was completed, 20 patients (9%) had at least one
HLA-A,B,DR identical donor. Mixed lymphocyte culture (MLC) tests were
performed for 20 patients, and 16 (5% of the initial searches) were

found to have at least one HLA-D compatible donor. Hopefully, this yield could be improved if larger numbers of donors were available.

The relationship between the estimated proportion of successful searches and donor registry size is illustrated in Figure 3. These calculations were based on HLA population data from three sources: the Ninth International Histocompatibility Workshop ("WS 1980"); the International Collaborative Renal Transplant Registry of G. Opelz ("ICTS, 1984"); and a panel of local blood donors from the Puget Sound Blood Center ("Local Blood Donors"). As expected, an increasing fraction of patients will find an HLA identical donor as the number of potential donors increases. Unfortunately, this correlation is not linear and, according to these estimations, even with $> 10^5$ donors only 75% of patients are expected to match. This finding is consistent with the theoretical prediction, based on the frequencies of the known HLA alleles and permulations of all possible combinations, that the maximum number of HLA-A,B,DR phenotypes is $> 10^6$.

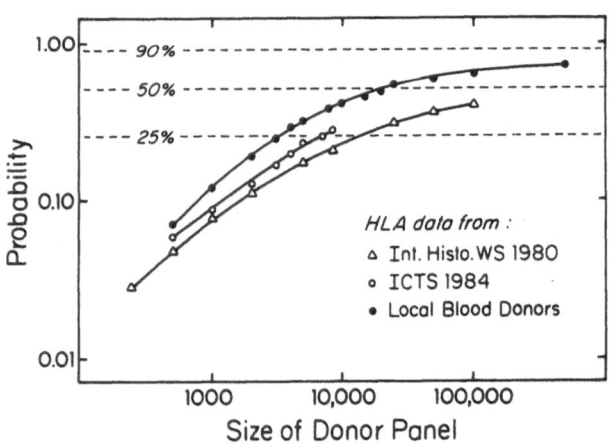

FIGURE 3.   PROBABILITY OF FINDING UNRELATED HLA-A,B,DR
            IDENTICAL DONORS

IV.   NEED FOR AN UNRELATED VOLUNTEER DONOR NETWORK

According to a survey by the International Bone Marrow Transplant Registry (IBMTR), a total of 2,398 allogeneic or syngeneic transplants were performed worldwide in 1984 by 162 marrow transplant teams (38). This compares to less than 1,000 transplants performed in 1981. It has been estimated that every year in the U.S.A. there are approximately 12,000 patients per year in the U.S.A. who could potentially benefit from a marrow transplant but do not have an HLA identical sibling (39). Before we have the capability of providing transplants for all these patients, we will need much larger numbers of HLA typed volunteer marrow donors.

To recruit and HLA type a sufficiently large number of volunteer marrow donors, some kind of inter-regional cooperation is necessary. Individual Donor Centers are unlikely to have the resources to establish donor systems of more than 10^3 to 10^4 individuals. Combining donor files from several centers can provide a larger panel having a genetically more diverse composition. Essential responsibilities of a Donor Registry (NBMDR) will be directed to coordinating efficient communications between Transplant Centers and Donor Centers and establishing appropriate standards for protecting the safety and confidentiality of the volunteer donors.

A National Unrelated Bone Marrow Donor Network is being established in the U.S.A. through the joint efforts of the American Red Cross (ARC), American Association of Blood Banks (AABB), and the Council of Community Blood Centers (CCBC). Initial funding is provided by a federal contract grant from the Office of Naval Research. In the United Kingdom, an independent marrow donor system with more than 50,000 volunteers has been established by the Anthony Nolan Foundation, and in France an effort is underway to recruit and type volunteers for a national marrow donor registry. Discussions about developing similar donor systems are taking place in several other countries.

Multiple searches for unrelated marrow donors have been carried out between individual Transplant Centers and Donor Centers located in different countries. Communication and arrangements, however, become increasingly complex across international boundaries. Eventually, formal agreements will be necessary in order to enable a more efficient access to potential matched donors worldwide.

References

1.   Thomas ED, Storb R, Clift RA, Fefer A, Johnson FL, Neiman PE,
Lerner KG, Glucksberg H, Buckner CD (1975).   Bone marrow
transplantation.   N Engl J Med 292:832-843, 395-902.

2.   Thomas, ED (1983). Marrow transplantation for malignant diseases.
J Clin Oncol 9:517-531.

3.   Elkins WL (1971).   Cellular immunology and the pathogenesis of
graft-versus-host reaction.   Prog Allergy 15:178-187.

4.   Hansen JA, Woodruff JM, Good RA (1981).   The graft-vs-host
reaction in man: Genetics, clinical features and immunopathology.   In:
Immunodermatology Vol. 7.   Safai B, Good RA (eds).   New York, Plenum,
pp 229-257.

5.   Thomas ED, Lochte HL, Cannon JH, Sahler DD, Ferrebbe JW (1959).
Supralethal whole body irradiation and isologous marrow
transplantation in man.   J Clin Invest 38:1709-1716.

6.   Bortin MM, Rimm A (1977).   Severe combined immunodeficiency:
Characterization of the disease and results of transplantation.   JAMA
238:591-600.

7.   Good RA, Bach FH (1974).   Bone marrow and thymus transplants:
Cellular engineering to correct primary immunodeficiency. In: Clinical
Immunobiology, Vol. 2. Bach FH, Good RA (eds). New York, Academic
Press, pp 63-114.

8.   Reisner Y, Kapoor N, Kirkpatrick D, Pollack MS, Cunningham-
Rundles S, Dupont B, Hodes MZ, Good RA, O'Reilly RJ (1983).
Transplantation for severe combined immunodeficiency with HLA-A,B,D,DR
incompatible parental marrow cells fractionated by soybean agglutinin
and sheep red blood cells.   Blood 61:341-348.

9.   Storb R, Thomas ED, Buckner CD, Appelbaum FR, Clift RA, Deeg HJ,
Doney K, Hansen JA, Prentice RL, Sanders JE, Stewart P. Sullivan KM,
Witherspoon RP (1984).   Marrow transplantation for aplastic anemia.
Semin Hematol 21:27-35.

10.   Bortin MM, Gale RP, Rimm AA, for the Advisory Committee of the
International Bone Marrow Transplant Registry (1981).   Allogendic bone
marrow transplantation for 144 patients with severe aplastic anemia.
JAMA 245:1132-1139.

11.   Anasetti C, Doney KC, Storb R, Meyers JD, Farewell VT, Buckner
CD, Appelbaum FR, Sullivan KM, Clift RA, Deeg HJ, Fefer A, Martin PJ,
Singer JW, Sanders JE, Stewart PS, Witherspoon RP, Thomas ED (1986).
Marrow transplantation for severe aplastic anemia: Long-term outcome
in 50 "untransfused" patients.   Ann Intern Med 104:461-466.

12.   Kaplan EL and Meier P (1958).   Nonparametric estimation from
incomplete observations. J Am Statis Assoc 53:457-481.

13.   Thomas ED, Buckner CD, Sanders JE, Papayannopoulou T, Borgna-
Pignatti C, De Stefano P, Sullivan KM, Clift RA, Storb R (1982).
Marrow transplantation for thalassemia. Lancet 2:227-228.

14.   Gluckman E, Berger R, Dutreix J (1984).   Bone marrow transplantation for Fanconi anemia. Sem Hematol 21:20-26.

15.   Thomas ED, Buckner CD, Banaji M, Clift RA, Fefer A, Flournoy N, Goodell BW, Hickman RO, Lerner KG, Neiman PE, Sale GE, Sanders JE, Singer J, Stevens M, Storb R, Weiden PL (1977).   One hundred patients with acute leukemia treated by chemotherapy, total body irradiation, and allogeneic marrow transplantation. Blood 49:511-533.

16.   Thomas ED, Buckner CD, Clift RA, Fefer A, Johnson FL, Neiman PE, Sale GE, Sanders JE, Singer JW, Shulman H, Storb R, Weiden PL (1979). Marrow transplantation for acute nonlymphoblastic leukemia in first remission. N Engl J Med 301:597-599.

17.   Thomas ED, Clift RA, Fefer A, Appelbaum FR, Beatty P, Bensinger WI, Buckner CD, Cheever MA, Deeg HJ, Doney K, Flournoy N, Greenberg P, Hansen JA, Martin P, McGuffin R, Ramberg R, Sanders JE, Singer J, Stewart P, Storb R, Sullivan K, Weiden PL, Witherspoon R (1986). Marrow transplantation for the treatment of chronic myelogenous leukemia. Ann Intern Med 104:155-163.

18.   Goldman JM, Apperley JF, Jones L, Marcus R, Goolden AWG, Batchelor R, Hale G, Waldmann H, Reid CD, Hows J, Gordon-Smith E, Catovsky D, Galton DAG (1986).   Bone marrow transplantation for patients with chronic myeloid leukemia. N Engl J Med 314:202-207.

19.   Appelbaum FR, Dahlberg S, Thomas ED, Buckner CD, Cheever MA, Clift RA, Crowley J, Deeg HJ, Fefer A, Greenberg P, Kadin M, Smith W, Stewart P, Sullivan KM, Storb R, Weiden P (1984).   Bone marrow transplantation or chemotherapy after remission induction for adults with acute nonlymphoblastic leukemia: A prospective comparison. Ann Intern Med 101:581-588.

20.   Champlin RE, Ho WG, Gale RP, Winston D, Selch M, Mitsuyasu R, Lenarsky C, Elashoff R, Zighelboim J, Feig SA (1985).   Treatment of acute myelogenous leukemia: A prospective controlled trial of bone marrow transplantation versus consolidation chemotherapy. Ann Intern Med 102:285-291.

21.   Appelbaum FR, Thomas ED (1983). Review of the use of marrow transplantation in the treatment of non-Hodgkin's lymphoma. J Clin Oncol 1:440-447.

22.   Thomas ED, Cronkite EP (1980).   Radiation Injury. In: Harrison's Principles of Internal Medicine, 9th edition. Isselbacher KJ, Adams RD, Braunwald E, Petersdorf RG, Wilson JD (eds.). New York, McGraw-Hill Book Company, pp 941-945.

23.   Beatty PG, Clift RA, Mickelson EM, Nisperos B, Flournoy N, Martin PJ, Sanders JE, Storb R, Thomas ED, Hansen JA (1985).   Marrow transplantation from related donors other than HLA identical siblings. N Engl J Med 313:765-771.

24. Storb R, Deeg HJ, Thomas ED, Appelbaum FR,Buckner CD, Cheever MA, Clift RA, Doney KC, Flournoy N, Kennedy MS, Loughran TP, McGuffin RW, Sale GE, Sanders JE, Singer JW, Stewart PS, Sullivan KM, Witherspoon RP (1985). Marrow transplantation for chronic myelocytic leukemia: A controlled trial of cyclosporine versus methotrexate for prophylaxis of graft-versus-host disease. Blood 66:698-702.

25. Irle C, Deeg HJ, Buckner CD, Kennedy M, Clift R, Storb R, Appelbaum FR, Beatty P, Bensinger W, Doney K, Cheever M, Fefer A, Greenberg P, Hill R, Martin P, McGuffin R, Sanders J, Stewart P, Sullivan K, Witherspoon R, Thomas ED (1985). Marrow transplantation for leukemia following fractionated total body irradiation: A comparative trial of methotrexate and cyclosporine. Leuk Res 9:1255-1261.

26. Storb R, Deeg HJ, Whitehead J, appelbaum F, Beatty P, Bensinger W, Buckner CD, Clift R, Doney K, Farewell V, Hansen J, Hill R, Lum L, Martin P, McGuffin R, Sanders J, Stewart P, Sullivan K, Witherspoon R, Yee G, Thomas ED (1986). Methotrexate and cyclosporine compared with cyclosporine alone with prophylaxis of acute graft versus host disease after marrow transplantation for leukemia. N Engl J Med 314:729-735.

27. Storb R, Deeg HJ, Farewell V, Doney K, Appelbaum F, Beatty P, Bensinger W, Buckner CD, Clift R, Hansen J, Hill R, Longton G, Lum L, Martin P, McGuffin R, Sanders J, Singer J, Stewart P, Sullivan K, Witherspoon R, Thomas ED (1986). Marrow transplantation for severe aplastic anemia: Methotrexate alone compared to a combination of methorexate and cyclosporine for prevention of acute graft-versus-host disease. Blood 68:119-125.

28. Martin PJ, Hansen JA, Buckner CD, Sanders JE, Deeg HJ, Stewart P, Appelbaum FR, Clift R, Fefer A, Witherspoon RP, Kennedy MS, Sullivan KM, Flournoy N, Storb R, Thomas ED. Effects of in vitro depletion of T cells in HLA identical allogeneic marrow grafts (1985). Blood 66:664-672.

29. Waldmann H, Hale G, Cividalli G, Weshler Z, Manor D, Rachmilewitz EA, Polliak A, Or R, Weiss L, Samuel S, Brautbar C, Slavin S (1984). Elimination of graft-versus-host disease by in vitro depletion of alloreactive lymphocytes with a monoclonal rat anti-human lymphocyte antibody (Campath-1). Lancet 2:483-486.

30. Herve P, Plouvier E, Flesch M, Racadot E, Cahn JY, Bernard A, Goldstein G (1986). Successful acute and chronic GVHD prevention and engraftment in 38 patients receiving T cell-depleted BMT-Besancon experience. Bone Marrow Trans 1:92.

31. Patterson J, Prentice HG, Brenner MK, Gilmore M, Blacklock H, Hoffbrand AV, Janossy G, Skeggs D, Ivory K, Apperley J, Goldman J, Burnett A, Gribben J, Alcorn M, Pearson C, McVickers I, Hann I, Reid C, Wardle D, Bacigalupo A, Robertson AG (1985). Graft rejection following T cell depleted BMT. Br J Haematol 61:562.

32. Meyers JD, Thomas ED (1982). Infection complicating bone marrow transplantation. In: Rubin RH, Young LS (eds): <u>Clinical Approach to Infection in the Immunocompromised Host</u>. New York, Plenum, pp 507-551.

33. Witherspoon RP, Lum LG, Storb R (1986). Immunologic reconstitution after human marrow grafting. Semin Hematol 21:2-10.

34. Hansen JA, Clift RA, Mickelson EM, Nisperos B, Thomas ED (1981). Marrow transplantation from donors other than HLA identical siblings. Hum Immunol 1:31-40.

35. Hansen JA, Clift RA, Thomas ED, Buckner CD, Storb R, Giblett ER (1980). Transplantation of marrow from an unrelated donor to a patient with acute leukemia N Engl J Med 303:565-567.

36. Gordon-Smith EC, Fairhead SM, Chipping PM (1982). Bone marrow transplantation for severe aplastic anemia using histocompatible unrelated volunteer donors. Br Med J 285:835-837.

37. "Household and Family Characteristics," In: Current Population Reports, Series P-20, #411, March 1985, U.S. Bureau of Census.

38. Bortin MM, Rimm AA (1986). Increasing utilization of bone marrow transplantation. Transplantation 42:229-234.

39. Gale RP (1986). Potential utilization of a national HLA-typed donor pool for bone marrow transplantation. Transplantation 42:54-58.